# TRENDS IN COMPUTER AIDED INNOVATION

## IFIP – The International Federation for Information Processing

IFIP was founded in 1960 under the auspices of UNESCO, following the First World Computer Congress held in Paris the previous year. An umbrella organization for societies working in information processing, IFIP's aim is two-fold: to support information processing within its member countries and to encourage technology transfer to developing nations. As its mission statement clearly states,

> *IFIP's mission is to be the leading, truly international, apolitical organization which encourages and assists in the development, exploitation and application of information technology for the benefit of all people.*

IFIP is a non-profitmaking organization, run almost solely by 2500 volunteers. It operates through a number of technical committees, which organize events and publications. IFIP's events range from an international congress to local seminars, but the most important are:

- The IFIP World Computer Congress, held every second year;
- Open conferences;
- Working conferences.

The flagship event is the IFIP World Computer Congress, at which both invited and contributed papers are presented. Contributed papers are rigorously refereed and the rejection rate is high.

As with the Congress, participation in the open conferences is open to all and papers may be invited or submitted. Again, submitted papers are stringently refereed.

The working conferences are structured differently. They are usually run by a working group and attendance is small and by invitation only. Their purpose is to create an atmosphere conducive to innovation and development. Refereeing is less rigorous and papers are subjected to extensive group discussion.

Publications arising from IFIP events vary. The papers presented at the IFIP World Computer Congress and at open conferences are published as conference proceedings, while the results of the working conferences are often published as collections of selected and edited papers.

Any national society whose primary activity is in information may apply to become a full member of IFIP, although full membership is restricted to one society per country. Full members are entitled to vote at the annual General Assembly, National societies preferring a less committed involvement may apply for associate or corresponding membership. Associate members enjoy the same benefits as full members, but without voting rights. Corresponding members are not represented in IFIP bodies. Affiliated membership is open to non-national societies, and individual and honorary membership schemes are also offered.

# TRENDS IN COMPUTER AIDED INNOVATION

*Second IFIP Working Conference on Computer Aided Innovation, October 8-9 2007, Michigan, USA*

*Edited by*

**Noel León-Rovira**
*ITESM, México*

Springer

*Trends in Computer Aided Innovation*

Edited by N. León-Rovira

p. cm. (IFIP International Federation for Information Processing, a Springer Series in Computer Science)

ISSN: 1571-5736 / 1861-2288 (Internet)

**ISBN 978-1-4419-4530-3** eISBN: 978-0-387-75456-7
Printed on acid-free paper



9 8 7 6 5 4 3 2 1
springer.com

## Contents

# Preface

It is a great honor and pleasure to welcome you to Brighton on the occasion of the 2nd IFIP Working Conference on Computer Aided Innovation.

As we started with the first ideas of organizing an international conference on Computer Aided Innovation 5 years ago, we got different answers that went from skepticism up to curiosity and enthusiasm.
The 1st Conference took place in Ulm, Germany in November 2005 and the 2nd Working Conference Computer Aided Innovation in Brighton, Michigan is now a reality. More than 40 contributions from all over the world were received from which 23 have been selected for being published after a thorough reviewing process. The Working Group 5.4 Computer Aided Innovation has now 34 members form 13 different countries.

The design of products and systems is becoming more demanding and complex as the amount of new products and newly developed interdisciplinary technologies increase exponentially. The product and process development must now take into account the entire lifecycle while avoiding environmental damage and facilitating the use of new technologies and physical principles. A structured and goal-oriented innovation process is nowadays mandatory for innovation success. Therefore, innovation supporting software becomes a key factor for the innovation process that guides the project teams through the complexity of the market.

Computer aided simulations are becoming more acknowledged as efficiency and effectiveness of the innovation process is becoming a key competitive advantage. Computational simulations related to the compatibility of novel ideas introduced in a computational social group that suggest acceptance of rejection are being also researched. This new kind of simulations produces insights that provide innovators with new tools to enhance their performance and effectiveness.

The transition from resource-based products to knowledge-based products is compelling the New Product Development process to be more innovative and efficient, making innovation processes even more challenging. Methods for structural and topological optimization, based on generative algorithms, are now used by practitioners to obtain optimal geometrical solutions that were earlier only possible after costly and time consuming trial and error approaches. As Product Life Cycle Management is being integrated with Knowledge Management methods and tools, new alternatives arise regarding the creation of new paradigms of the Engineering Desktop. New Knowledge-Based Engineering systems support innovators' activity through rules and knowledge re-use, thus reducing the product development time while increasing its functionality, quality and reducing environmental damage.

Some existing ideas and concepts of Computer Aided Innovation (CAI) focus on assisting product designers in their creative stage, however a more comprehensive vision conceives more comprehensive CAI systems that begin at the creative stage of perceiving business opportunities and customer demands, that also offer support for developing inventions and also provide help for turning inventions into successful innovations in the market. It is now accepted that computers have an important role to play in helping innovators to find the direction and solutions for new products and processes. In order to achieve higher performance in product and process innovation the Theory of Inventive Problem Solving (TRIZ) is being integrated into CAI Systems. As a result new methods and tools are taking shape under the name of Computer Aided Innovation that are now helping engineers and technicians to find innovative solving approaches. CAI therefore stands out as being a break from the usual trends.

The 2nd IFIP Working Conference on Computer Aided Innovation will play an important role in clarifying the essential factors characterizing these new arising methods and tools for bridging the gap between the traditional methods and current trends in search of efficient innovation. The Conference provides a forum for presenting and discussing current research and recent advancements in all fields of supporting innovation with computer tools for product and process development.

The aims of the papers presented at the conference are:

- Making contributions for clarifying the role of computer aided innovation tools.

- Contributing to the further development of the Engineer's Desktop focusing on end-to-end product creation process with methods and tools to ensure the feasibility and success of innovations in early stages of the innovation process product-related validation (e.g. requirements processing, simulation) process-related validation (e.g. frontloading).
- Enhancing the engineering innovation activity with computer tools and methods.
- Discussing organizational, technological and cognitive aspects of the application of CAI methods and tools, and also contributing to the evaluation of their effectiveness and efficiency.
- Developing the theoretical foundations of CAI.

The contributions published in this proceeding represent the state of the art in research, development and implementation in these fields. They include:

- Research Papers describing contributions and latest results of fundamental investigations;
- Industrial Papers presenting applications of CAI experiences and methods;
- Speculative Papers advancing experiences with new theories, approaches and methods, without necessarily offering validated results;
- Engineering Creativity and Innovation Papers on new experiences from education, training, teamwork and case examples.

The contributions are focused on following topics:

- The innovation process;
- Engineering design and innovation;
- TRIZ, CAI and Artificial Intelligence;
- Basic definitions to CAI;
- Fundamental approaches: CAD/CAE and CAI software;
- Optimization and innovation;
- Shape and topology generators and optimization;
- Integration of CAI methods and tools into engineering processes;
- Innovation in collaborative networks of enterprises;
- Social impact of innovations.

We would like to extend our sincere thanks and appreciation to the International Program Committee for their reviewing efforts and to the Local Organizing Committee for their hard work toward making this Working Conference a success.

N. León-Rovira , Co-Chair 2nd IFIP Working Conference on Computer Aided Innovation
Director Research Chair Creativity, Inventiveness and Innovation
Center for Innovation in Products and Processes
Tecnológico de Monterrey, Campus Monterrey

## Contributors

Name(s)/Affiliations

**Tan Runhua, Zhang Jianhui, Yang Bojun, Tian Yumei, Liang Yanhong, Ma Jianhong, Lian Benning, Cao Guozhong, Bai Zhonghang, Zhang Peng;** Hebei University of Technology, Tianjin, P.R. China.

**Mikael Nybacka, Tobias Larsson, Åsa Ericson;** Luleå University of Technology, Sweden.

**Ricardo Sosa;** ITESM Campus Querétaro, Mexico.

**John S. Gero;** George Mason University, Fairfax, VA, USA and University of Technology, Sydney, Australia.

**Darrell Mann;** Systematic-Innovation, Cleavedon, United Kingdom

**Greg Yezersky;** Institute of Professional Innovators, Farmington Hills MI, USA.

**Udo Kannengiesser;** NICTA, Australia

**Hong Liu, Xiyu Liu;** Shandong Normal University, Jinan, China.

**Alexander Hesmer, Karl. A. Hribernik, Jannicke Baalsrud Hauge, Klaus-Dieter Thoben;** University of Bremen, Bremen, Germany.

**Gaetano Cascini, Davide Russo;** Università di Firenze, Florence, Italy.

**Manuel Zini;** DrWolf srl, Florence, Italy.

**Nikolai Khomenko, Roland DeGuio, Thomas Eltzer, Denis Cavallucci, François Rousselot;** INSA Strasbourg Graduate School of Science and Technology - LgéCo, Strasbourg Cedex, France.

**Noel León, Jose Cueva, Cesar Villarreal, Sergio Hutron, German Campero, Humberto Aguayo;** ITESM Campus Monterrey, Mexico.

**Umberto Cugini, Marco Ugolotti;** Politecnico di Milano, Milano, Italy.

**Albert Albers, Thomas Maier;** University of Karlsruhe, Germany.

**Tomasz Arciszewski, Kenneth A. Shelton;** George Mason University, Fairfax, VA, USA.

**Seung-Hyun Yoo;** University of Maryland, Maryland, USA, On leave from Ajou University, Suwon, Korea.

**Eung-Jun Park, Jae-Sil Lee, Joon-Ho Song, Dae-Jin Oh, Woong-Rak Chung, Yeongtae Lee, and Dhaneshwar Mishra:** Ajou University, Suwon, Korea.

# UXDs-Driven Contradiction Solving For Conceptual Design Using CAIs

Tan Runhua
Institute of Design for Innovation, Hebei University of Technology,
Tianjin, 300130, P.R. China
Email: rhtan@hebut.edu.cn

**Abstract.** Design is situated, which means the explicit consideration of the state of environment, the knowledge and experience of the designer and the interaction between the designer and the environment during designing. Central to the notion of situated design is the notion of design situation and constructive memory. When Computer-aided innovation systems (CAIs) are applied in the design, the environment and the situation are different from the traditional design process and environment. The basic principles of some CAIs in the world market are directly related to theory of inventive problem solving (TRIZ). Special TRIZ solutions, such as 40 inventive principles and the related cases, are medium-solutions to domain problems. The second stage analogy process is used to generate domain solutions and in this process the TRIZ solutions are used as source designs of analogy-based process. Unexpected discoveries (UXDs) are the key factors to trigger designers to generate new ideas for domain solutions. The type of UXDs for the specific TRIZ solutions is studied and an UXDs-driven contradiction solving for conceptual design is formed. A case study shows the application of the process.

**Keywords.** unexpected discovery, Contradiction solving, Conceptual design, computer-aided innovation

## 1 Introduction

A design situation models a particular state of interaction between a design agent or a designer and the environment at a particular point in time [1]. Memory construction occurs whenever a design agent or a designer uses past experiences and knowledge within the current design environment in a situated manner [2]. When computer aided innovation systems (CAIs) [3] are applied in design process the design situation becomes specific. The interactions mainly happen among designers and a serial of interfaces produced from CAIs.

The basic theory or method of some important existing CAIs [3] is TRIZ, that is, theory of inventive problem solving [4]. TRIZ is developed by analyzing and

---

*Please use the following format when citing this chapter:*

Runhua, T., 2007, in IFIP International Federation for Information Processing, Volume 250, Trends in Computer Aided Innovation, ed. León-Rovira, N., (Boston: Springer), pp. 1-11.

inducing a great deal of patents within the world database. TRIZ has put forward the concepts, models and tools of inventive problem solving. But the TRIZ special solutions, such as selecting forty or less inventive principles from all the forties, are not domain solutions needed by designers. It's still a problem how to convert the TRIZ special solutions into domain solutions when TRIZ is applied.

One of the converting processes is an analogy-based process [5]. Unexpected discoveries (UXDs) [6] are the key factors to trigger designers to generate new ideas for domain solutions during this process [5]. How to find UXDs from the TRIZ special solutions is becoming an important step for obtaining the domain solutions.

There are a few inventive problems to be solved in TRIZ, such as technological maturity mapping, technology evolution, function solving, contradiction solving etc. But the contradiction solving is the most important kind of problems to be solved and it is rooted philosophy in TRIZ [7]. This study will be restricted to find solutions for this kind of inventive problems. The types of UXDs from the TRIZ special solutions will be studied. And an UXDs-driven analogy based process for conceptual design will be formed.

## 2 Design situation for contradiction solving using CAIs

The development of different CAIs based on TRIZ has made the TRIZ more applicable and practical. There are one or more knowledge bases in CAIs, in which much knowledge is abstracted from the world patent bases. The knowledge is arranged by the framework of TRIZ.

Fig. 1, which shows contradiction solving method in TRIZ, is developed for explaining the principle of CAIs. There are two parts in the figure, TRIZ world and outside world. In the outside world, the designers find a domain technical contradiction for a domain problem and input it into the TRIZ world. In TRIZ world the contradiction is firstly to be transformed into a standard contradiction using 39 engineering parameters in TRIZ and then forty or less inventive principles are selected through the matrix. Also, some design cases following the principles are contained. The selection of these principles and the cases with these principles are TRIZ special solutions. The design cases are the results of analyzing patent bases from outside world. The TRIZ world in Fig. 1 has been programmed as a kind of arithmetic and a module of CAIs, such as in the Goldfire Innovator and InventionTool, which is then developed. Interaction between TRIZ world and outside world is realized by the interfaces of the CAIs.

The knowledge, which is tacit in different domains of a patent base, is difficult to be applied by designers because it is a problem to find a useful one in different domain. If a patent abstracted in any domain is stored in the case base of TRIZ it becomes explicit knowledge or codified knowledge which can be easily found and applied for idea generation. In the knowledge base of CAIs, a case is described using a sketch with text to explain the working principle of that sketch. When one principle as a TRIZ special solution is selected all the cases relevant to that principle can be browsed one by one. New ideas for the domain solutions may be formed from designers' mind during the browsing process. Fig. 2 shows the model of design situation for this specific environment.

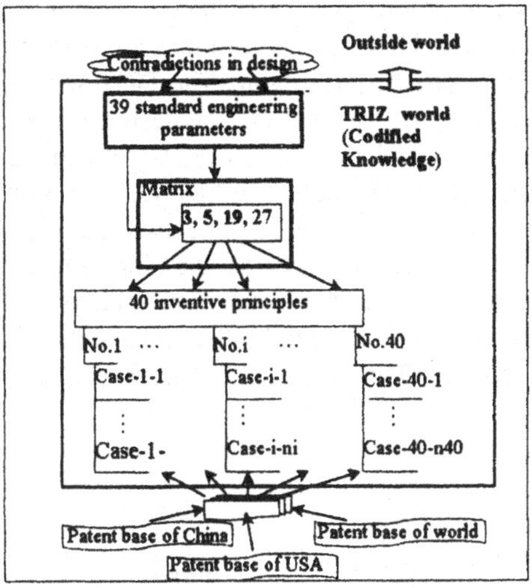


**Fig. 1.** Contradiction solving model in TRIZ

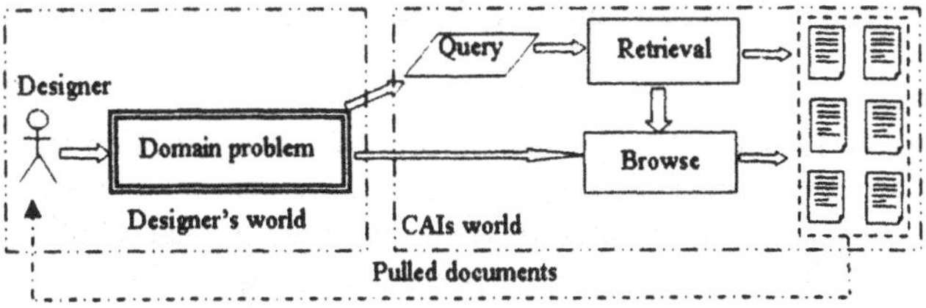


**Fig. 2.** Design situation using CAIs for technical contradiction

In the designer's internal world, the selected principles and cases as codified knowledge are apperceived, and some cues are found, which trigger for designers to form domain solutions. In Fig. 2, the creative ability for designers, that is, ability to produce ideas, should be increased.

# 3 A macro-analogy-based process Using CAIs

Analogies are partial similarities between different situations that support further inference. Analogy-based design (ABD) means the application of analogy to design. The ABD is used not only for the normal traditional design, but also for the innovative and creative design [8]. In the process of ABD, the existing designs and the designs to be carried out are source designs and goal designs respectively. One of the conditions to carry out ABD is the existence of source or base designs in different domains in large number.

Designers find domain problem and then convert them into TRIZ problems, such as contradictions. Then, the TRIZ special solutions are determined from TRIZ, such as using the matrix for contradiction solving. The TRIZ special solutions are not the domain solutions needed by designers. Last, the designers themselves convert TRIZ special solutions into domain solutions. There are two mappings in this process, from domain problem (DP) to TRIZ special solution (TSS) and from TRIZ special solution to domain solution (DS). Both of the two mapping process are analogy processes. The first mapping process is called the first stage analogy process (FSAP) and the second mapping is called the second stage analogy process (SSAP) [5]. The first stage analogy process is completed by the application of CAIs and the outputs are TRIZ special solutions. The second stage analogy process is a human-based process.

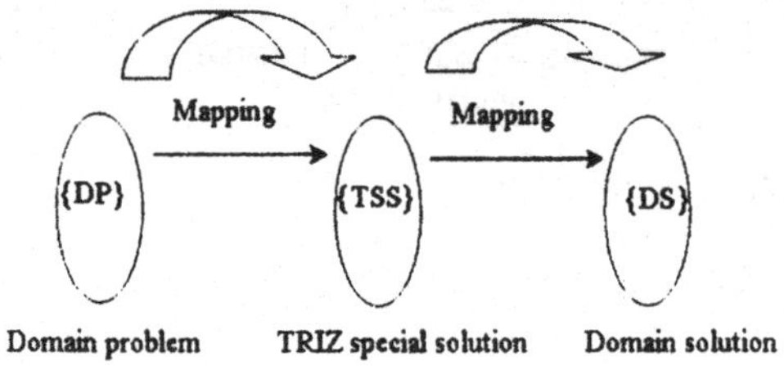


**Fig. 3.** Three domains of the design world using TRIZ

Suwa and Gero [9] have developed a concept "situated-invention (S-invention)", which means a designer generates the issue or requirement for the first time in the current design task in a way situated in the design setting. Gero et al. have studied the generation of S-invention, and summarized a design process [6, 8, 10, 11]. Firstly, the design agents apperceive the domain problem and determine source design and goal design. Then design agents find unexpected discoveries (UXDs) through the matching of source design and goal design. UXDs are transferred to goal design by mapping, and new goal is generated. Then modified goal design is produced. There may be multi-source designs, and the last modified goal design is the concept of solving domain problems through modifying goal design continually.

Gero and his group are major in architectural design. So the source designs are drawings of different kinds of architectures. If the source designs are substituted by TRIZ special solutions and the cases corresponding to them the design process for generation of S-invention can be applied to generate the domain solutions. Designers find several UXDs and modify goal design depending on their design experience, the comprehension of domain problems and the situation. At last some modified goal designs are domain solutions. The macro-process of ABD for contradiction solving using TRIZ is shown in Fig. 4.

The contradiction analogs are the standard contradictions selected from the 39 engineering parameters, which have similar meanings to the domain contradictions. The UXDs about solving contradiction analogs are found from the principles solving contradiction analogs and cases having solved contradiction analogs.

The main processes of implementation for the process of Fig. 4 are how to find UXDs about solving the contraction analogs and convert these UXDs into ideas for solving the domain contradictions. UXDs enlighten designers on invention and make new concepts or ideas appear, so discovering and transferring UXDs are the key to the success. According to the concept of constructive memory [2], the memory is not direct reappearance of former experience but a function of former experience, which changes after producing these experience and situation of memory requirement. An UXD is a "new" perceptual action that has a dependency on "old" physical action(s) [9]. This means that if a designer traces or pays attention to the existence of source designs, the perceptual action is an instance of UXD. Experiences and UXDs drive designers to generate new concepts or ideas. The new concepts are mapped to the goal design to produce a new goal design.

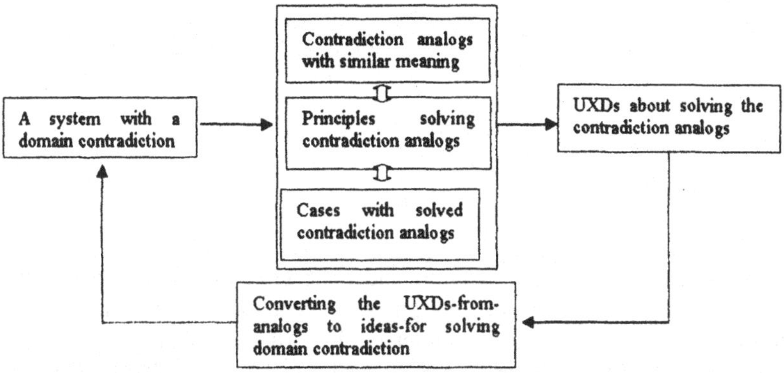


**Fig. 4.** The macro-process of ABD for contradiction solving using

Perceptual actions [9] are operated for architectural drawings. They must be extended to TRIZ special solutions and cases for TRIZ based design.

## 4 Types of UXDs for contradiction solving

Currently, there are several well-known design theories and methodologies developed for general (macro-scaled) systems, such as Systematic Design Methodology [12], Axiomatic Design Theory [13] etc. Essentially, the design knowledge representation in these theories is based on the Function-Behavior-Structure (FBS) model [14-17]. There are different kinds of functions, behaviors and structures [18-21]. Functions are divided into atomic, source, destination and transfer functions. And behaviors are divided into three kinds, continuous-time-behavior, discrete-time-behavior and state-transition-behavior. According to the specific design context, a structure may refer to a sub-system, a sub-assembly, a component, a feature, or a geometric entity, and a physical relationship.

For new designs there will be a transition from functional model to a structure model at some point during design process. The transition is called mapping. There are two kinds of mapping, function-structure mapping and function-behavior-structure mapping. The former is suitable for the mapping of extrinsic functions and the latter is suitable for intrinsic functions.

In order to make new ideas for innovation, the designers need to apply the knowledge of different domains, especially with which the designers are unfamiliar. Generally, the first step for latent in any problem solving is the definition of a space in which the solution of the problem is believed to exist. This solution space is shaped through the requirements, restrictions and constraints imposed on the situation by the problem-solver interpreting user and company needs and functional, structural and intentional. When TRIZ is applied, TRIZ special solutions are sources to extend solution spaces including extended function spaces, behavior spaces and structure spaces, as shown in Fig. 5.

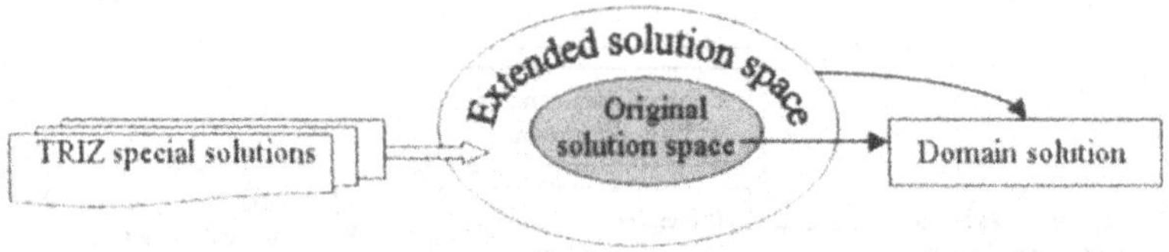


**Fig. 5.** Extended solution space

Designers generate domain solutions under the search in original and extended solution spaces. The core cues to designers are functions, behaviors or structures implicated in the TRIZ special solutions. The original solution space that the designers have known is used as a background for the domain solutions. The reason is that the designers could not generate new solutions using the original solution space that they have known.

Suwa and Gero [9] have devided UXDs into three types, depending on what types of visuo-spatial feature that the designer discovers. One is the discovery of a visual feature such as shape, size or texture of a previously-drawn element. The second is the discovery of a spatial or organizational relation among more than one previously-drawn element. The third is the discovery of a space that exists in previously drawn elements. The types are suitable for design of an architecture, in which the basic elements are dots, lines, rectangles, circles, arrows and so on. For complex system design, such as complex mechanical system design, the basic elements are more complex. The types divided are not suitable for them. New types are needed.

Because function, behavior and structure are basic knowledge representations and they are also major UXDs implicated in the TRIZ special solutions. For contradiction solving using TRIZ and CAIs, the UXDs are divided into four types that designers discover from the CAI interfaces. Table 1 shows the types, definition of each type, and the instances of each type and how to find an UXD.

**Table 1.** Types of UXDs

| Types | Definition | Instances | How to find |
|---|---|---|---|
| UXD-1 | An inventive principle which is suitable for solving the contradiction faced | One to four inventive principles | Check the matrix using two engineering parameters |
| UXD-2 | A function that one case implicated | atomic, source, destination and transfer functions | Physical actions |
| UXD-3 | A behavior that one case implicated | continuous-time-behavior, discrete-time-behavior, state-transition-behavior | Physical actions |
| UXD-4 | A structure that one case implicated | A sub-assembly, a component, a feature, or a geometric entity, and a physical relationship, | Physical actions |

For the designers using CAIs, the physical actions to find a UXD are only looking, which means designers looking at the computer screen: the principles of contradiction solving and the cases showing by pictures and contexts, shown in Fig. 2. By looking actions designers discover the UXDs implied.

# 5 UXDs driven analogy based design using TRIZ

A new idea is not direct reappearance of a designer's former experience but a function of former experience, changes after producing these experience and situation of memory requirement. A new idea is a memory constructed by some stimulus from the situation faced by designers. After the first stage of analogy process using TRIZ four or less inventive principles and related cases are at hand. Designers analyze the principles and cases through physical actions and find some UXDs of different kinds. Now the situations faced by designers are these UXDs. The designers construct new memories under stimulus of UXDs, and as result a domain solution is constructed suddenly. Fig. 6 shows the process which is driven by UXDs analogy process.

According to the Fig. 6, the process is divided into 7 steps, which are as following:

Step 1: Identify domain problem. Analyze social needs, customer needs or products existed and identify domain problems in the form of contradictions.

Step 2: Find TRIZ special solutions. Under the situation CAIs find principles and relevant cases.

Step 3: Identify primary goal design. It can be formed from the existed products.

Step 4: Find UXDs. The physical actions are performed for the principles and cases.

Step 5: Transfer UXDs to goal designs. Generate constructive memories and modify goal designs.

Step 6: Make a judgment. If modified goal designs do not satisfy the needs or the numbers of satisfied goal designs are scarce, turn to step 4, whereas the step continues.

Step 7: Following design: evaluation, embodiment and detail design.

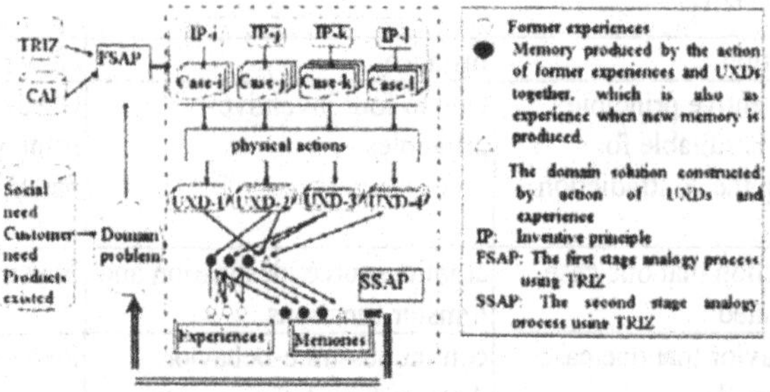


**Fig. 6.** A process of UXDs driven analogy based design

The step 2 belongs to the first stage analogy process and the step 3 to 6 belongs to the second stage analogy process. The evaluation is included in step 7, and many methods have been developed for this application. The result of the second stage analogy process may be a domain solution or multi-solutions. For the multi-solutions, it needs to determine one or a few domain solutions through evaluation. The result of detailed design is design files, drawings or data files, which are applied for manufacturing.

## 6 Case study

Dropping pills, produced by dropping pill machines, are a kind of Chinese traditional medicine. After dropping and drying, they should be put into little bottles for selling in the market. The machine to put the pills into bottles is a kind of package machines. There are no standard machines of this kind. A few machines have been developed by one or two firms in China. But new principles for the kind machines are needed by firms of medicine production.

The main functions of the machine are distributing pills, discharging pills, bottling and lidding. Here, the concentration will be on the ideas generation for structure of discharging pills.

Step 1: Identify domain problem. The function of distributing bills is implemented by a structure of tumbling cylinder shown in Fig. 7. The pills are distributed in the several circles. The implementation of discharging pills should be based on the principle of distributing bills. Fig. 8 shows a principle for discharging pills being currently used. If the roller is redesigned and used inside the cylinder a transmission mechanism is also needed and that is not easy. So a contradiction is between adaptability (No.35) and complexity of a device (No.36).

Step 2: Find TRIZ special solutions. InventionTool3.0, which is CAIs, is applied in this step. There is a model in the system which is contradiction solving. Select the improved parameter 'adaptability' and worse parameter 'complexity of a device', then, the interfaces show the TRIZ special solutions, which are No.29(Pneumatic or hydraulic construction), No.15(Dynamicity), No. 28(Replacement of mechanical system), No. 37(Thermal expansion). The four principles and the relevant cases in the case base are the TRIZ special solutions.

Step 3: Identify primary goal design. The Fig. 7 is identified as the starting point.

Step 4: Find UXDs. By browsing the cases the designer may find several UXDs and generate several ideas under the stimulants of UXDs. Fig. 8 is a case which shows the principle of a component used in a kitchen machine. When working, the ball moves up and down and releases the exhausted gas produced during frying. The principle implies an UXD, which is the ball moves under the pressure of exhausted gas. The UXD is a kind of behavior which is an UXD-3 in table 1.

Step 5: Transfer UXDs to goal designs. Convert the UXD into the new ideas and generate a principle for discharging pills. The pill moves under the pressure of air flow if a pill is the ball of Fig. 9. A possible structure for this principle is shown in Fig. 10.

Step 6: Make a judgment. More ideas may generated by finding more UXDs. The structure shown in Fig. 10 is usable.

Step 7: Following design. Fig. 11 is the conceptual design in which the four main functions of the machine are implemented.

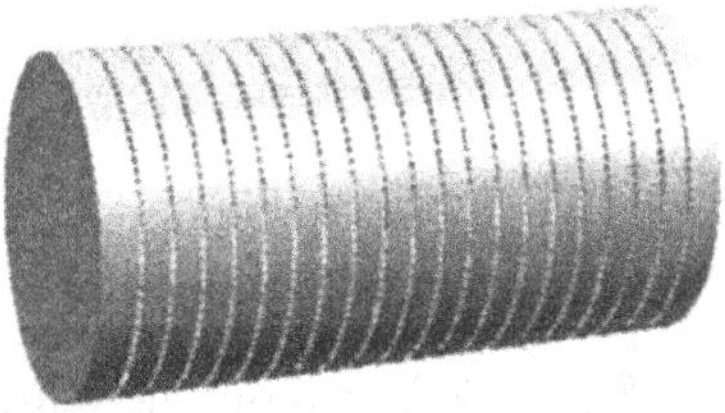

**Fig. 7.** The cylinder for distributing pills

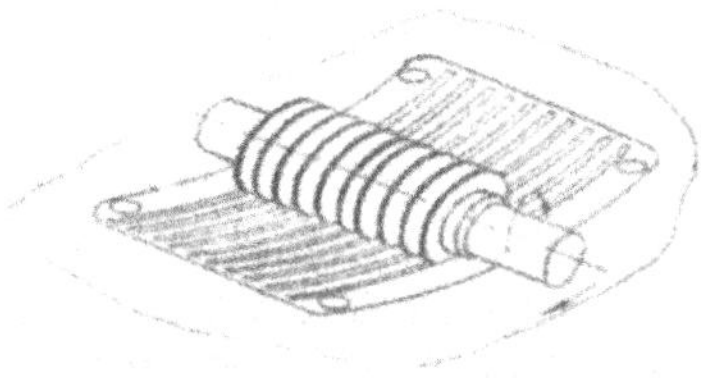

**Fig. 8.** A principle for discharging pills

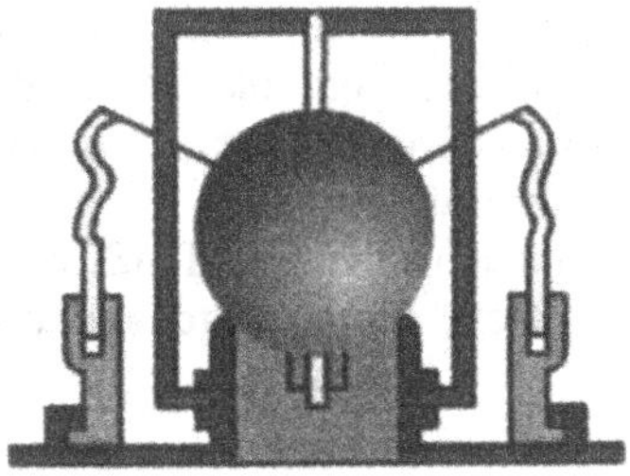

**Fig. 9.** A case in No.15

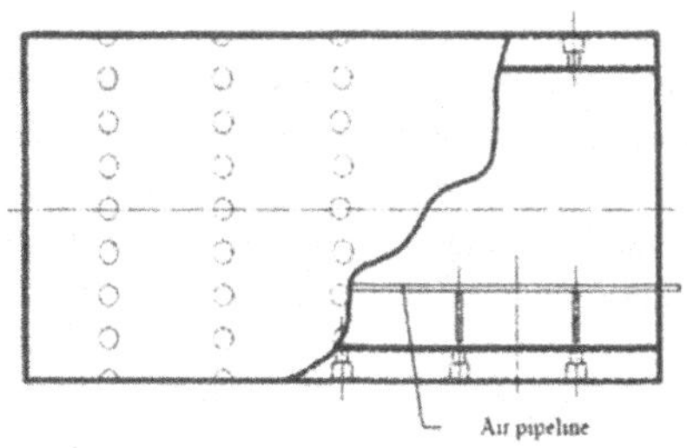


**Fig. 10.** Structure for discharging pills

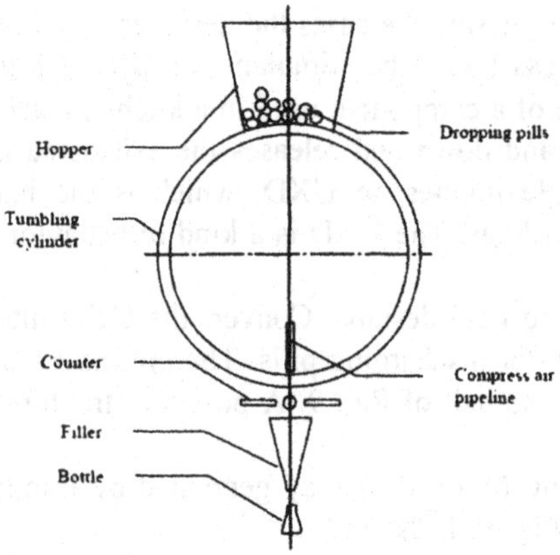


**Fig. 11.** A conceptual design of packaging machine for dropping pills

## 7 Conclusions

CAIs show all inventive principles and cases, which are the source designs of ABD. The databases of CAIs are a fruit of TRIZ researchers for many years, which have broadly applicability. The application of the database will improve the validity of ABD that has been extensively accepted by designers.

When TRIZ is applied to solve a contradiction in design the first and second stage analogy process are existed. The results of the first stage analogy process are source designs of the second stage analogy process. To find UXDs from the sources is the key step to generate successful ideas for innovation. Four types UXDs have been divided. The physical action for finding UXDs from the computer screen of CAIs is only looking.

A seven step process model is formed for design, in which UXDs are driving force for generating new ideas. Designers find UXDs from the TRIZ special solutions and react with experiences that designers have to construct memories suddenly. Then new ideas for domain solutions are formed.

The model put forward is only related to contradiction solving of TRIZ. It needs to extend the model to technological evolution, effects, and standard solutions of TRIZ in order to effective application of CAIs.

## 8 Acknowledgement

The research is supported in part by the Chinese Natural Science Foundation under Grant Numbers 50675059 and Chinese national 863 planning project under Grant Number 2006AA04Z109. No part of this paper represents the views and opinions of any of the sponsors mentioned above.

## 9 References


1. G.J. Smith and J.S. Gero, What Does an Artificial Design Agent Mean by Being Situated?, *Design Studies*, 26, 535-561 (2005).
2. J.S. Gero, Constructive Memory in Design Thinking, *In: Design Thinking Research Symposium: Design Representation*, Goldschmidt G and Porter W, eds. MIT, Cambridge, 29-35 (1999).
3. S. Kohn, S. Husig and A. Kolyla, Development of an Empirical Based Categorsation Scheme for CAI Software, In:*1st IFIP TC-5 Working Conference on CAI*, ULM, Germany, Nov.14-15, 143-157 (2005).
4. G. Altshuller, The Innovation Algorithm, TRIZ, Systematic Innovation and Technical Creativity, Worcester, *Technical Innovation Center*, INC (1999).
5. R. Tan, Process of Two Stages Analogy-based Design Employing TRIZ, *International Journal of Product Development*, 4(1/2), 109-121 (2007).
6. J. Kulinski and J.S. Gero, Constructive Representation in Situated Analogy in Design, In: *CAAD Futures 2001*, Kluwer, Dordrecht, Vries B and Achten H, eds, 507-520 (2001).
7. G.M. Martin, How Combinations of TRIZ Tools are Used in Companies – Results of a Cluster Analysis, *R&D Management*, 35( 3), 285-296 (2005).
8. J.S. Gero, Concept Formation in Design: Towards a Loosely Wired Brain Model, Candy L and Hori K. eds. In: *Strategic Knowledge and Concept Formation Workshop*, Loughborough University of Technology, 135-146 (1997).
9. M. Suwa and J.S. Gero and T. Purcell, Unexpected Discoveries and Inventions of Design Requirements, *Design Studies*, 21, 539-567 (2000).
10. S. Ricardo and J.S. Gero, Computational Models of Creative Situations, Gero J S and Brazier FMT, eds. In: *Agents in Design 2002 Key Centre of Design Computing and Cognition*, University of Sydney, 165-180 (2002).
11. J.S. Gero, Design Tools as Situated Agents that Adapt to Their Use, Dokonal W and Hirschberg U, eds. In: *eCAADe21, eCAADe*, Graz University of Technology, 177-180 (2003).
12. G. Pahl and W. Beitz, *Engineering Design – a Systematic approach*, 2nd Edition, London, Springer (1996).
13. N.P. Suh, *Axiomatic Design- Advance and Application*, New York, Oxford University Press (2001).
14. A. Chakrabarti and T.P. Bligh, An Approach to Functional Synthesis of Mechanical Design Concepts: Theory, Application, and Emerging Research Issues, *AIEDAM: Artificial Intelligence for Engineering Design, Analysis and Manufacturing*, 10(5), 313-332 (1996).
15. L. Qian and J.S. Gero, Function-behavior-structure Paths and Their Role in Analogy-based Design, *Artificial Intelligence for Engineering Design, Analysis Manufacturing*, 10(4), 289–312 (1996).
16. X.F. Zha and H. Du, Mechanical Systems and Assemblies Modeling using Knowledge Intensive, *Artificial Intelligence for Engineering Design, Analysis and Manufacturing (AIEDAM)*, 15(2), 145-171(2001).
17. G. Cao and R. Tan, FBES Model for Product Conceptual Design, *International Journal of Product Development*, 4(1/2), 22-36 (2007).
18. F. Zhang and D. Xue, Distributed Database and Knowledge Base Modeling for Concurrent Design, *Comput-Aided Design*, 34(1), 27–40 (2002).
19. D. Xue and H. Yang, A Concurrent Engineering-oriented Design Database Representation Model, *Computer-Aided Design*, 36, 947–965 (2004).
20. R.B. Stone, and K.L. Wood, Development of a Functional Basis for Design, Transactions of the ASME, *Journal of Mechanical Design*, 122(4), 359–370 (2000).
21. A.K. Goel, Design, Analogy and Creativity, *IEEE Expert*, 12(3), 62-70 (1997).

# Computational Explorations of Compatibility and Innovation

Ricardo Sosa[1] and John S. Gero[2]
1 Department of Industrial Design, ITESM Querétaro, Mexico.
rdsosam@itesm.mx
2 Krasnow Institute for Advanced Study and Volgenau School of Information Technology and Engineering, George Mason University, USA.
john@johngero.com
Web page: http://www.computationalcreativity.com/

**Abstract.**. This paper presents a range of computational simulations related to the compatibility of novel ideas that suggest interesting phenomena regarding divergence and convergence, social influence and patterns of change. These computational studies produce insights providing the researcher with another tool to reason about these challenging problems. According to current theory, innovations that are perceived by social groups as having greater compatibility will be adopted more rapidly than other innovations. However, compatibility plays a role in some of the paradoxes of creativity and innovation and its real implications in a range of situations remain unclear.

**Keywords.** Computational social simulation, compatibility, innovative design.

## 1 Introduction

Computational social simulations have been developed to address a number of questions on the link between creativity and innovation [1]. The focus of these studies has been the interaction of the individual level of agency (the change agent) with its social and cultural context in the processes of generation and evaluation of new ideas. Generation and evaluation are regarded as complementary in the dyad novelty-utility found in the canonical definition of creativity. The term "creative situations" captures this assumed coupling or alignment at two levels of agency, namely the individual factors and the contextual conditions, resulting in:

a) a match between individual attributes and actions within the appropriate context in order to generate novel ideas, and
b) the relevant environmental processes that facilitate diffusion, adoption and advantageous consequences of innovations.

---

*Please use the following format when citing this chapter:*

Sosa, R., Gero, J. S., 2007, in IFIP International Federation for Information Processing, Volume 250, Trends in Computer Aided Innovation, ed. León-Rovira, N., (Boston: Springer), pp. 13-22.

In a) a range of individual and micro-level factors are involved: the preparation and expertise of different individuals, their various abilities to perceive and adequately formulate problems, their access to positions that enable implementation of and experimentation with ideas, initial support for diffusion, etc. This range of factors is directly related to the generative phase of novel ideas. In b) a number of social and macro-level factors are involved: a corpus of predecessors' achievements, information dissemination channels, social perceptions of problems, solutions and drawbacks, norms and practices, production and distribution infrastructure, cultural constraints, etc. This range of factors is directly related to the evaluative phase of novel ideas by a social group.

Whilst factors in a) can be associated with "logic and genius" in Simonton's model of creativity [2], b) provides a level to capture processes of "chance and zeitgeist" in that same model. The concept of creative situations [1] proposes that a type of alignment is necessary between these levels of agency to enable the generation and the evaluation of creative and innovative ideas. Computational social simulations provide useful means to grow these types of theoretical constructs [3].

## 1.1 Compatibility

According to current theory, innovations that are perceived by social groups as "having greater relative advantage, compatibility, trialability, observability, and less complexity will be adopted more rapidly than other innovations" [4]. Compatibility is an interesting topic of study because it plays a role in some of the paradoxes of creativity and innovation [1]:

- Original ideas may require freedom at many levels, yet constraints – such as compatibility with previous solutions and infrastructure – can actually benefit creativity and innovation [2, 5].
- The adoption of new ideas tends to increase as they mature. High quality and commercial success are usually found not in radical innovative ideas, but in more compatible modifications, such as "second generation" products [6].
- High compatibility may cause technological innovations to be more successful (higher adoption and diffusion degrees), yet the opposite may be true for artistic innovations, where the expectation is to break away from current standards [4]. What level of compatibility would better predict the success of a new design idea?

This paper addresses the relationship between compatibility and innovation via computational social simulations, aimed at clarifying or reformulating these apparent contradictions. A model is presented based on the DIFI framework of creativity [7, 8] and the FBS design prototype schema [9]. Our integrative framework is based on the complementarity of generative and evaluative processes by individuals and groups in design [1, 10].

Compatibility is conceptually defined here as the degree to which a novel idea shares attributes or properties with dominant or competing ideas. This can be expressed in several ways in any given computational implementation, for instance if the designs generated and evaluated by agents are represented numerically, their compatibility can be calculated by their shared numerical attributes. If the designs are represented by geometrical shapes, compatibility can be given by their shared geometrical attributes, possibly as they are perceived by a group of agents.

## 2 Framework

This section describes the conceptual architecture of our framework rather than the technical implementation details which can be found elsewhere [1]. This enables us to place a stronger emphasis here on the types of hypotheses embedded in the framework, the types of experiments carried on these studies, and the types of results obtained.

Computational social simulation refers to the study of social agency through the ideation, implementation, and execution of computer models usually built under rather simple assumptions with which the experimenter is able to define a series of hypotheses and formally implement and experiment with them to explore the consequences of their interaction over time.

The type of computational systems that we have built in recent years have centred on the idea of social groups (implemented as multi-agent systems or cellular automata) whose members interact in order to generate and evaluate a range of ideas. 'Ideas' can be represented here by numeric values or geometrical shapes, and agent behaviour involves the exchange of values or perceptions of shapes between agents. This enables the modelling of societies where some agents aim to introduce novel ideas that are subsequently valued by their social groups.

In simple models (i.e., cellular automata), an explanatory limit of causality is quickly reached, given that randomness importantly influences the generation of values and their dissemination in constrained spaces of interaction, i.e. typically two-dimensional rectangular grids. Interesting variations include experimentation with other types of spaces, but a rather more useful approach involves modelling "bigger" agents in rich social spaces. Typically this means that randomness is replaced by a more grounded approach to guide the processes of generation and evaluation of ideas.

We have thus implemented multi-agent systems where ideas are represented as two-dimensional geometrical shapes that some agents (designers) generate and the rest (societies) perceive, evaluate and ultimately adopt or reject. This is implemented by individual mechanisms of shape perception including geometrical properties like boundaries, number of sides, angles, and transformations like uniform and non-uniform scale, rotation, etc.

The social spaces where agents interact is also enriched by including mechanisms of social influence in various dimensions: societies converge and diverge over time re-shaping groups of agents that share preferences, perceptions and/or decisions regarding existing ideas. Figure 1 shows the system architecture used in this paper as a framework to study compatibility with three main interacting system elements: the individual agent (designer), a social evaluative group (field), and their environment or domain [1, 10].

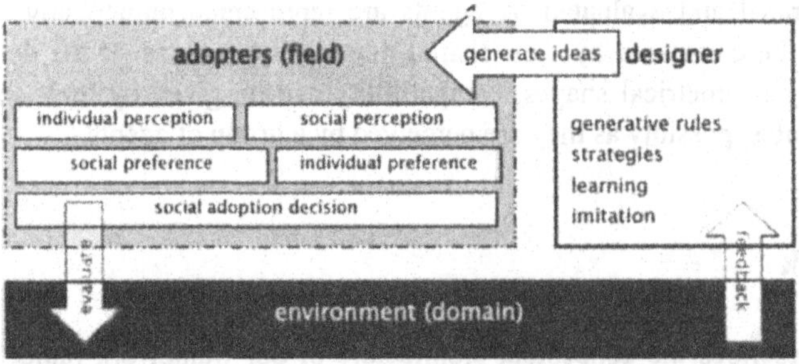


**Fig. 1.** Graphic description of the framework with three interacting elements: the individual agent (designer), a social evaluative group (field), and environment (domain).

## 2.1 Domain-Field-Individual

In this framework, the domain represents the set of values or ideas shared by a field. It typically includes the competing ideas in a social group at time $t_i$ as well as a cumulative number of ideas selected by the society during the simulation, $t_{0<i}$. In our multi-agent models the domain is usually implemented as a dynamic array where successful ideas are stored, possibly with a rate of decay representing the lifespan of the 'collective memory' of a society. Inclusion of ideas into the domain is implemented by a bottom-up mechanism by which agents that influence others gradually gain authority until a few of them exert the role of 'gatekeepers' of the domain. We have explored a range of possible mechanisms observing different emerging patterns of gatekeeping [10].

The field is defined in this framework by the aggregate characteristics of the agents and their interaction over time in different social spaces. In each of these spaces, cycles of convergence and divergence can be 'grown' as in Axelrod's classical model of influence [11]. The adoption decisions can be constrained by the confluence of these social spaces. The implementation of a social space can consist of running a cycle regulating all agent interactions by a given criterion. In a social space of preferences, agents may exchange or influence each other's bias towards certain geometrical features in their adoption decisions; in a space of perceptions, agents may exchange or influence each other's attributes of competing shapes. Experimental settings here include modifying the rate of exchange at different spaces, the interaction rules, and the type of data structure used in the implementation.

Lastly, the individual agent is defined in this framework by the set of design rules carried by the agents that besides evaluation are able to generate new or modify existing shapes, which are subsequently available for social evaluation. Here the range of experimental settings is large and includes generative mechanisms, competition strategies, novelty seeking motivation, distributions of traits and abilities, rates of creation, etc. The role of the individual agent can be implemented via an evolutionary system, analogy making, case-based reasoning or any other generative process potentially including direct human participation, although we have not yet explored this hybrid approach. An implementation of this framework can make use of a geometrical shape representation that captures some of the properties of design solutions. Moreover, this representation supports reasoning mechanisms for adoption decisions based on the geometrical properties of sets of two-dimensional line representations constrained by 12 boundary points as shown in Figure 2(a). This is a simple way of representing features of design ideas with nomological constraints.

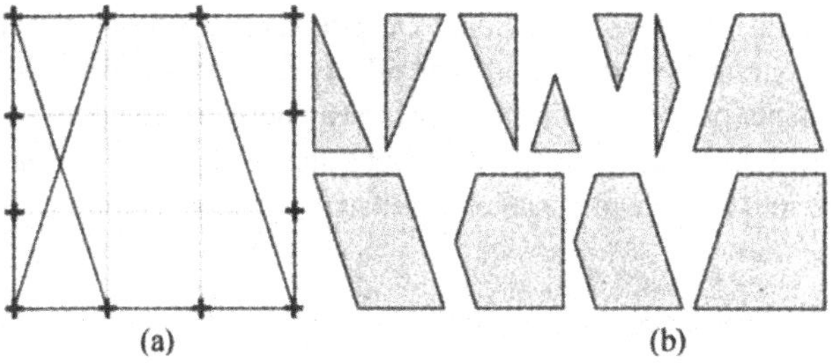

**Fig. 2.** (a) A simple geometrical shape perception and (b) some possible interpretations built by different adopters based on individual perception biases.

Multiple representation and ambiguity are possible because ideas are perceived and interpreted by adopters according to a set of randomly distributed perception biases. Figure 2(b) shows sample perceived features of an idea. The assumption is that people perceive design ideas in (marginally) different ways and therefore base their evaluations on different features of those ideas. By manipulating experimental variables at the domain, field and individual levels, we are able to explore in our computational models the formation of patterns over simulated time of social influence, diffusion, and emergence of new values. We have commenced by manipulating a few variables independently, registering their effects and assessing the framework's ability to capture phenomena observed in field and laboratory studies published in the literature [1].

### 2.2 A Simulation Run

The role of designers is modelled here as instances of change agents that work towards providing novel solutions to a set of problems shared by large social groups. Typically, in these social simulations a small set of up to half a dozen designers compete by iteratively interpreting the problem and proposing a solution which is evaluated by the whole social group including other designers. The designer agents learn from the feedback provided by the social group including their adoption decisions and a measure of satisfaction with their adopted solutions. Designer agents also have a learning mechanism that influences their future behaviour based on the overt actions of their competitors and the social adoption of their solutions. Although the evaluation process carried by adopter agents follows a set of rules that define individual perception and preferences (following a normal distribution), social interaction is included as the potential of adopter agents to influence each other's decisions to adopt or reject solutions generated by the designers.

Three social spaces are implemented in the studies reported in this paper: a space where geometrical preferences are exchanged, a second space where shape perceptions are exchanged, and a third space where adoptions decisions are exchanged. In social groups of a few hundred adopters, patterns of interest arise such as the emergence of opinion leaders and cycles of convergent-divergent adoption. During a simulation, the system is set to track the behaviour of every agent as well as the global patterns of group behaviour. Despite their apparent simplicity, these models of co-evolution generate non-linear effects that emerge from the interaction of their components over time. In this way, researchers are equipped with in silico laboratories where they can 'grow up' different states from a set of initial conditions, gaining insights into the role of designers as change agents in complex systems.

### 2.3 Compatibility Studies

In simple cellular automata models of social influence, compatibility has been identified as a key determinant of interaction [11]. Global group convergence tends to emerge as the aggregate effect of distributed local exchanges based on the gradual development of regions of compatible values. Starting from random conditions, the group tends to converge in one dominant value or reach a lock-in state where regions of incompatible values emerge. These systems have been extensively replicated showing that the final outcome of group convergence is highly likely depending on key variables such as the range of values assigned and the rules of interaction between neighbours or adjacent cells. These variations determine the likelihood of compatibility between cells in the grid and between regions of cells. If adjacent regions in a cellular automata develop compatible values, it is inevitable that a single dominant value will emerge either by dominance of one region over the others, or by combination of their compatible values. If incompatibility occurs, interaction is halted across regions and global convergence is not reached.

In a multi-agent system implementation of the framework presented in this paper where domain, field and individual design agents interact, it is possible to inspect the concept of compatibility further. Using an idea representation like the one described earlier based on geometrical shapes, compatibility can be measured as a degree of similarity. Figures 3(a) to 3(d) show a range of shape perceptions with different indices of compatibility based on shared geometrical characteristics. Figure 3(a) and 3(b) are more compatible since they share seven line segments, whilst Figure 3(a) and 3(c) are less compatible since they have only three line segments in common. In the same vein, Figures 3(b), 3(c) and 3(d) are compatible because they share symmetric properties, whilst Figure 3(a) is incompatible symmetry-wise. Likewise, Figures 3(a) and 3(b) are compatible in that they both present right angles, whilst Figures 3(c) and 3(d) do not.

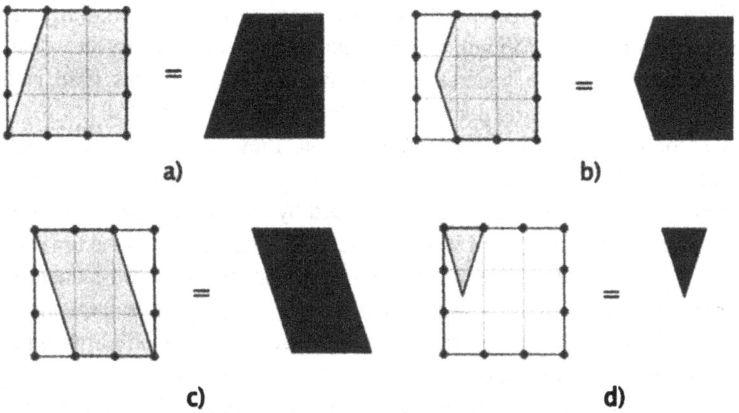


**Fig. 3.** Compatibility between ideas is implemented here based on the shared geometrical properties of shapes. Number of line segments, angles, symmetry, and other properties can be incorporated by evaluating agents to determine the degree of compatibility in any set of shapes.

Compatibility can evolve during a simulation run for any agent in relation to any given shape due to agent interaction in different social spaces. This is possible due to the constantly evolving preferences, perceptions and adoption decisions being continuously exchanged in the social group and periodically updated due to the introduction of new ideas by the design agents.

The following are the experimental settings explored in this paper:

1. Conditions are first explored in relation to compatibility and adoption of new ideas. Monte Carlo simulations traverse the idea compatibility space. This is implemented by running simulations with identical initial conditions in all control variables except the generative processes of designers, which are manipulated to generate new ideas that go from entirely incompatible to entirely compatible, namely $0 < c < 1$, where $c$ is the degree of compatibility as estimated by the designer agent introducing the idea into the system. This corresponds to one extremum where new ideas are entirely random to the other where new ideas are identical to existing ideas (at least from the designer's viewpoint, which can be marginally different as measured by some social groups which operate with varying perceptions). The results of the system at the three levels are recorded on every simulation run for every initial condition (results represent an average over ten runs for every step in the parameter space). The field effects of compatibility is analysed, i.e., the adoption patterns.
2. A slight variation addresses the effect that compatibility has on the type of innovation observed in a society. The assumption is that, with all other conditions kept constant, changes in compatibility of new ideas may yield an output of either radical or transformational innovations in the system. In line with the literature, radical innovations are characterised by the qualitative differences between succeeding dominant ideas in a society. Low levels of compatibility may yield radical innovations, whilst marginal differences may

emerge from generative processes that promote high compatibility. This is implemented by Monte Carlo simulations traversing the space of compatibility in the generative process. The focus in this case is in the analysis of the resulting domain.

3. A second aspect of interest is competition. Previous studies have suggested that the rate of the generative processes of new ideas may have a non-linear effect on innovation [1]. This is explained by a "glass ceiling" that imposes a limit on the frequency and scope of cycles of change, due to the time required for new ideas to be disseminated. It is not clear what could be the effects of compatibility and rate of generation of new ideas. This is implemented by Monte Carlo simulations traversing the spaces of compatibility and rate of behaviour by designer agents.
4. A third question is addressed regarding compatibility and complexity. Theories suggest that innovations are more effective if new ideas are more compatible and less complex [4]. Our framework enables experimentation by traversing the compatibility and complexity spaces of new ideas in the generative processes. As in the previous settings, all other conditions are kept constant, whilst the generative processes of designer agents in the system are controlled at the initial time enabling analysis on the outputs at domain and field levels when a) compatibility is low and complexity is high, b) compatibility is high and complexity is low, c) both are high, and d) both are low. Complexity in this framework can be measured by the length of the representation of the geometrical shapes. One key assumption here is that more complex shapes will enable a higher diversity of perceptions of new ideas by members of the social group.

### 2.4 Results

A set of key implications result from our simulations related to the compatibility of novel ideas introduced in a social group.

1. Low levels of compatibility may yield high levels of divergence in a social group, causing information flow to stop and thus, precluding innovation. This has been characterised in equivalent modelling approaches as the emergence of "few or many distinct cultural regions depending on the scope of cultural possibilities, the range of interactions, and the size of the geographic territory" [11].

2. High levels of compatibility may cause total and rapid convergence in a social group. Whilst constant cycles of change take place under such conditions, the impact of novel yet highly compatible ideas is minimal. Namely, such simulations typically show continuous cycles of 'transformational' innovations, i.e., where small variations of a dominant idea are repeatedly introduced.

3. If novel ideas with low levels of compatibility are introduced in a social group, but information flow is sustained during long time periods (externally or otherwise), a high rate of crossover of ideas is likely. In such cases, periodical cycles of change have large impacts (changes are significant and have large scope). In addition, such cases show that the 'culture' of a society may change radically even if social structure remains unchanged.

4. Low levels of compatibility may yield opportunistic innovations if the rate of idea production is high enough to support a competitive environment. An opportunistic innovation is defined here as the wide adoption of a new idea that draws attributes from competing new ideas, maintaining their advantage but increasing their compatibility.

5. A general consensus in the literature is that high compatibility combined with low complexity yield relatively fast diffusion rates and a reasonable scope of diffusion [4]. In our framework, less complex designs are those that can be represented with a smaller range of attributes, numerical, geometrical or otherwise. Our simulations illustrate that compatibility and complexity may exhibit undesirable effects given that in solutions with very low levels of complexity, a small attribute variation between two designs can rapidly decrease their compatibility. In contrast, high levels of complexity support a large variance of solutions marginally differentiated and thus, solutions with high levels of compatibility. Therefore, a balance between high compatibility and low complexity may be hard to achieve, accounting for their exceptional joint occurrence.

## 3 Discussion

A key potential implication of these studies is that isolated characteristics of designers and their ideas are insufficient to formulate conclusions about creativity and innovation. Causality may rather be inspected in the situational factors that define the relationship between designers and their evaluators. This framework enables the study of compatibility and innovation from a situational viewpoint, suggesting ways in which key characteristics of innovations may have very different causes and consequences depending on the surrounding contextual conditions. The following design guidelines can be formulated:

1. Design solutions must be perceived as having an adequate degree of compatibility with previous or competing alternatives.
2. In designing innovative solutions, the likely rate of diffusion must be estimated in order to adjust the degree of compatibility to avoid rapid, unstable and difficult to control flows that prevent assimilation of novel ideas.
3. The degree of compatibility of novel ideas may determine the extent to which novel ideas are reinterpreted or combined with existing dominant ideas. In some cases it may be desirable to allow for crossover, whilst in other cases (i.e., intellectual property) this may need to be avoided. This may be addressed by the relation between complexity and compatibility of novel and old ideas.

The computational exploration of compatibility and its interplay with complex phenomena like creativity and innovation yields promising results. Emergence is a key aspect to understand phenomena such as "creative situations". The results are not easily predictable, neither are they definite or necessarily valid against external conditions. Rather, these studies provide insights that provide the researcher with another tool to reason about these challenging problems. One way to advance this research methodology would be to contrast these findings with documented cases in the literature, and as aids to design experimental settings in the laboratory or the field.

## 4 References


1. R. Sosa, Computational Explorations of Creativity and Innovation in Design, *PhD Thesis*, Key Centre of Design Computing and Cognition (University of Sydney: Sydney, 2005).
2. DK. Simonton, *Creativity in Science: Chance, Logic, Genius, and Zeitgeist* (Cambridge University Press: Cambridge, 2004).
3. JM. Epstein, *Generative Social Science: Studies in Agent-Based Computational Modeling* (Princeton University Press: New Jersey, 2007).
4. EM. Rogers, *Diffusion of Innovations* (The Free Press: New York, 1995).
5. D. Partridge, and J. Rowe, *Computers and Creativity* (Intellect: Oxford, 1994).
6. H. Petroski, *The Evolution of Useful Things* (Knopf: New York, 1992).
7. M. Csikszentmihalyi, in: *The Nature of Creativity, Contemporary Psychological Perspectives*, edited by RJ Sternberg (Cambridge University Press, 1988), pp. 325-339.
8. DH. Feldman, M. Csikszentmihalyi, and H. Gardner, *Changing the World: A Framework for the Study of Creativity* (Praeger: Westport, 1994).
9. JS. Gero, Design Prototypes. A Knowledge Representation Schema for Design, *AI Magazine*. Volume 11, Number 4, pp. 26-36 (1990).
10. R. Sosa, and JS. Gero, in: *Computational and Cognitive Models of Creative Design VI*, edited by JS. Gero, and ML. Maher (University of Sydney: Sydney, 2005), pp. 19-44.
11. R. Axelrod, The Dissemination of Culture: A Model with Local Convergence and Global Polarization, *The Journal of Conflict Resolution*. Volume 41, 2, pp. 203-226 (1997).

# Emergent contradictions: A synthesis of TRIZ and complex systems theory

Darrell Mann
5A Yeo-Bank Business Park, Kenn Road, Cleavedon, BS21 6UW, United Kingdom
Darrell.mann@systematic-innovation.com

**Abstract:** The paper focuses on the identification and resolution of conflicts and contradictions in complex systems. By constructing simple, computer-based bottom-up models of an exemplar complex system, we show the potential for the emergence of multi-phase macro-level outcomes. We then show how these discontinuous phase-shifts may be modelled as contradictions, and from there, how TRIZ tools can be used to generate innovative solutions.

**Keywords:** complexity, emergence, phase-shift, discontinuity, bottom-up

*"If you show someone their future, they have no future.*
*If you take away the mystery, you take away hope."*
Philip Dick, Paycheck [1]

*"Don't get so far ahead of the parade no-one knows you're in it."*
John Naisbitt [2]

## 1 Introduction

What do we mean when we use the word 'innovation'? Opinions vary, but consistent themes include novelty, the addition of value and profit. In this paper we will use a somewhat different definition in order to make an important distinction with 'optimization'. The distinction is especially needed in the computing environment, because while computers are extremely effective at doing the optimization task, they remain to all intents and purposes useless at performing innovation related tasks. Therefore the important distinction between the two things is that 'optimization' involves the identification of optimum balanced values of a known set of continuously variable parameters, whereas 'innovation' involves the presence of some kind of discontinuity. As shown in Figure 1, 'optimization' stays

*Please use the following format when citing this chapter:*

Mann, D., 2007, in IFIP International Federation for Information Processing, Volume 250, Trends in Computer Aided Innovation, ed. León-Rovira, N., (Boston: Springer), pp. 23-32.

within the current paradigm while 'innovation' involves some kind of discontinuous shift (however large or small) to another paradigm. This paper is about the latter. We already know that computers are better than humans at optimization. This is because continuous variables are amenable to mathematical modelling. Discontinuous shifts are generally speaking not.

However, the theme of this paper is that computers already have a profoundly important role to play in helping designers to find those discontinuities. In many technical systems, designers are readily able to find such discontinuities (trade-offs) without the use of a computer. Yet in systems that are fundamentally complex the discontinuities are often hidden from view. This paper examines how in these situations the construction of bottom-up system models and the use of multi-agent, programmable modelling environments can be used not only to identify discontinuities, but to provide designers with a more complete understanding of emergent system level behaviour. It is hoped that in this way, more robust, contradiction-breaking solutions may be identified. Thus we believe that computers already have a significant role to play – one in which they are already far superior to humans - in the innovation arena.

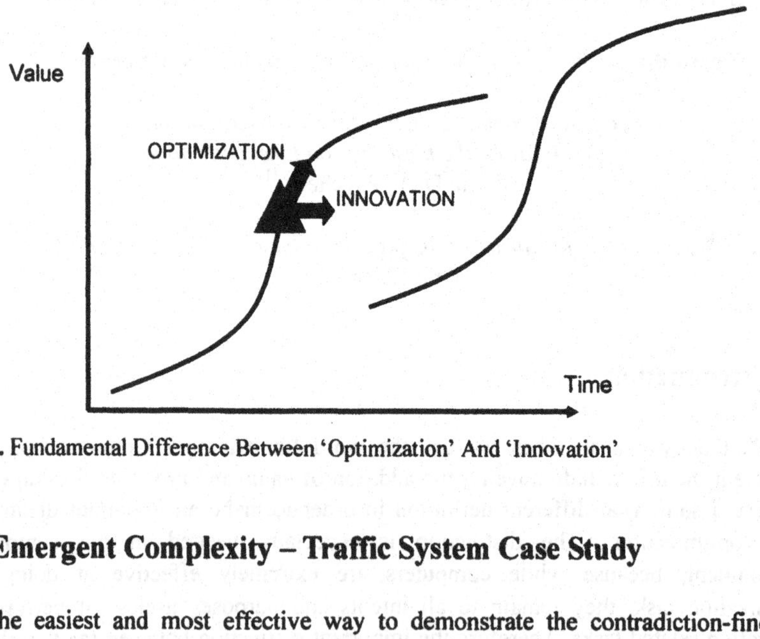


**Fig. 1.** Fundamental Difference Between 'Optimization' And 'Innovation'

## 2 Emergent Complexity – Traffic System Case Study

The easiest and most effective way to demonstrate the contradiction-finding abilities of computer software is to look at an example. The problem of highway traffic has probably been over-used, but we will examine it anyway. By using TRIZ we might generate one or two new insights into the problem. The traffic problem we will consider is flow and congestion on highways. Several researchers (Reference 3 for example) have sought to model highway traffic flow using multi-agent software models. The main idea here has been to demonstrate that often unexpected complex system-level behaviours will emerge from the combined effects of players (drivers in this case) who can possess no more than local knowledge.

In Figure 2 we describe some of those 'local' rules as they may apply to an individual driver inside their individual car. In most prior traffic flow study simulations it is assumed that every driver observes these local rules. In our simulations, we have expanded the analysis somewhat by identifying two distinctly different types of driver – one calm and relaxed; the other in a hurry and therefore stressed.

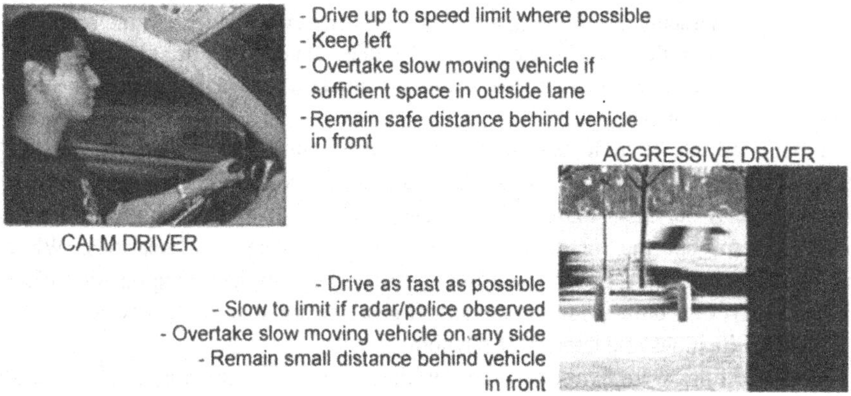

- Drive up to speed limit where possible
- Keep left
- Overtake slow moving vehicle if sufficient space in outside lane
- Remain safe distance behind vehicle in front
CALM DRIVER
AGGRESSIVE DRIVER
- Drive as fast as possible
- Slow to limit if radar/police observed
- Overtake slow moving vehicle on any side
- Remain small distance behind vehicle in front


**Fig. 2.** Defining 'Driver DNA' For Multi-Agent Modeller

While it is clear from other studies that 'all-drivers-follow-the-same-rules' models generate complex behaviour already, one of the aims in this work has been to try and establish the role that driver stress-level might have on emergent traffic behaviour. The simulation models that we built assumed that all drivers defaulted to the 'calm' state. The switch from 'calm' to 'aggressive' behaviour was programmed as a sudden shift in keeping with the 'leaky integrator' model of human brain function (Reference 4). In this model, transitions from one state to another occur abruptly once the level of a chemical messenger reaches a threshold level. Thus, the simulation attempts to model what the world knows about the way the brain operates. When subjected to stressful situations, the body generates more 'stress messengers'. The more stress, the more messengers get produced. At the same time, in order to prevent being drowned in a sea of chemicals, the body gradually disposes of the messengers. Here, then, is the 'leaky' part of the integrator story. An analogy would be filling a container with a small hole in the bottom. If more (stress messenger) fluid is put into the container than is able to drain through the small hole, then the container gradually fills up. If we fill it sufficiently, then eventually the container will become completely full and then start to overflow. This 'over-flow' situation is analogous to the sudden shift from 'calm' to 'aggressive' in the driver model. This is a part of the model that we know mimics reality quite closely. What is less clear is at what point in a highway driving situation drivers will shift from one state to another. In order to model this uncertainty, two stress-generator conditions were incorporated:

1. Stress messengers begin to be generated after drivers have been exposed to stationary traffic for more than a set period of time (this

period was modelled using a Monte Carlo randomization algorithm in order to reflect the belief that across a population, some people are become stressed more quickly than others); once chemical production has commenced, it will continue each time the driver is stationary after the first occasion.

2. Again controlled by a Monte Carlo algorithm to account for variation across the population, drivers were 'injected' with high rates of stress messengers following any kind of unexpected incident on the road. For example, if another driver braked suddenly or switched lane unexpectedly then this would be modelled as a rapid rise in stress messengers.

It is important to note that during the construction we had little quantified information against which to validate any part of the model. This would be wholly unacceptable if our task was to try and 'optimize' any part of the highway traffic flow problem, but the approach is valid for identifying non-linearities. While incorrect quantifications might result in us getting things in the wrong position if we tried to plot results on an absolute numerical grid, we are able to identify and model the relative situations and non-linear phenomena.

Once the rules were defined we ran multiple multi-agent software (Reference 5) simulations. Of primary interest during these simulations was the speed vehicles were capable of travelling during different traffic density conditions. Figure 3 shows a typical output when the simulations are used to plot the distance travelled along the highway by a succession of vehicles plotted as a function of time. What we should be looking for ideally in these plots is a succession of lines with constant gradients as in the left hand picture (gradient here being representative of speed).

Without going in to the specific details of what is shown in the right-hand Figure, what we actually see is very different. This is a plot describing the motion of many cars. The sharp ridges in the plots are indicative of sudden changes in speed of the vehicles now that the overall traffic density has exceeded a certain threshold value.

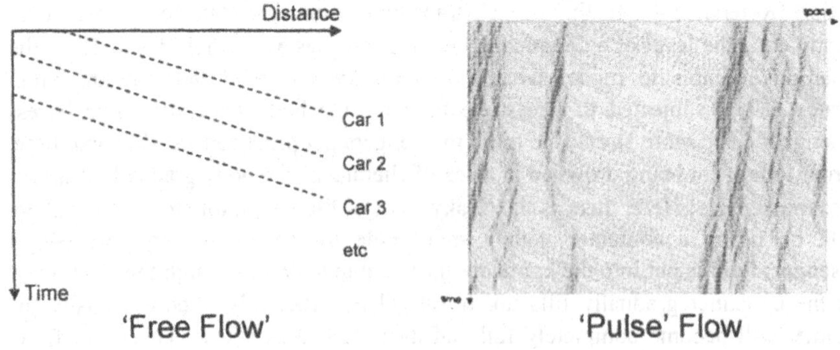


**Fig. 3.** Emerging Discontinuous Phase Transitions

In other words drivers all following the same set of linear rules can create a system behavior that is highly non-linear. If we analyze all of these non-linearities in detail, what we discover is that there are a number of very distinct modes of highway traffic flow. We can see these different modes plotted in Figure 4.

The reader may recognize some of these modes from their own personal highway driving experiences:

1. 'free-flow' traffic – in this mode, an individual driver is able to proceed unhindered by the effects of other traffic.
2. 'knotted-flow' – once the traffic density passes a certain value, the occasional slow vehicle on the highway tends to impede the progress of other vehicles. These 'knots' mean that other drivers are temporarily impeded in their progress until such times as they are able to access overtaking lanes and get past the slow moving impedance.
3. 'pulse' flow – this is perhaps the strangest phase of the four. The traffic speed oscillates, often wildly, between low and normal speeds; one second vehicles are driving normally, the next everyone is braking, then a few seconds later, vehicles rapidly accelerate up to normal speed again.
4. 'choked' flow – the most frustrating mode. Drivers find themselves stationary for periods, followed by periods of crawling progress.

The important thing to note here is that these four modes are discontinuously different from one another, undergoing a rapid transition from one mode to another. A useful analogy to keep in mind is the discontinuous phase transition that occurs when water is chilled to below zero degrees; one second it is liquid, the next ice. The transition from one mode to another occurs suddenly and with little or often no warning. The transitions offer the first seeds of opportunity for defining good contradictions to solve. However, it is worth first exploring in a little more detail how and why these phase transitions occur.

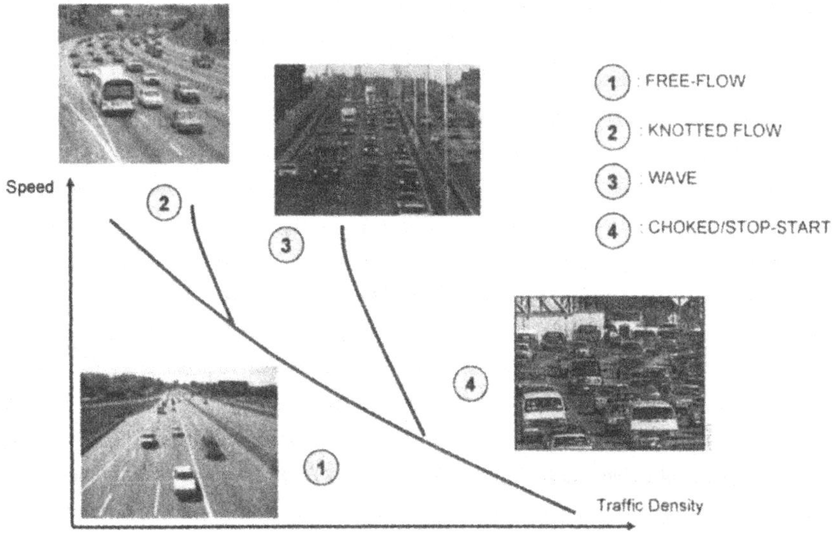


**Fig. 4.** Phase Transitions In Highway Traffic Flow

Emphasizing again the crucial distinction between optimization and innovation, the values of temperature, pressure, etc at which water makes the abrupt transition to ice are accurately known because thousands of experiments have been conducted. Yet even if we didn't know the precise freezing temperature of water, it would still be possible to clearly see the non-linear transition from liquid to solid taking place. In the case of the traffic simulation presented here, no experimental validation has taken place and so there is no possibility to state the actual speeds and traffic densities that denote the transition from one mode to another. Hence there has been no attempt to include numbers on the axes of the graph. Our interest rests with innovation, and the working hypothesis is that here the job is first and foremost about identifying the existence of non-linear transitions. Once the innovative solution concepts have been developed following the identification of the non-linear problems, we can worry about quantification and optimization.

## 2.1 Sub- And Super-Critical Phenomena

In the same way that it is possible to cool water below its zero degree 'critical point' and for it still to be a liquid, it is eminently possible for traffic to be travelling in one phase mode even though the speed-versus-density conditions indicate that it should be in another. In physics this condition is called 'super-critical'. Given the right conditions, water, can be cooled several degrees below zero and it still will not freeze. As soon as this 'super-critical' condition experiences a sudden perturbation (for example the presence of a tiny solid particle in the water acts as an ice formation initiation point), the whole system will rapidly switch to ice.

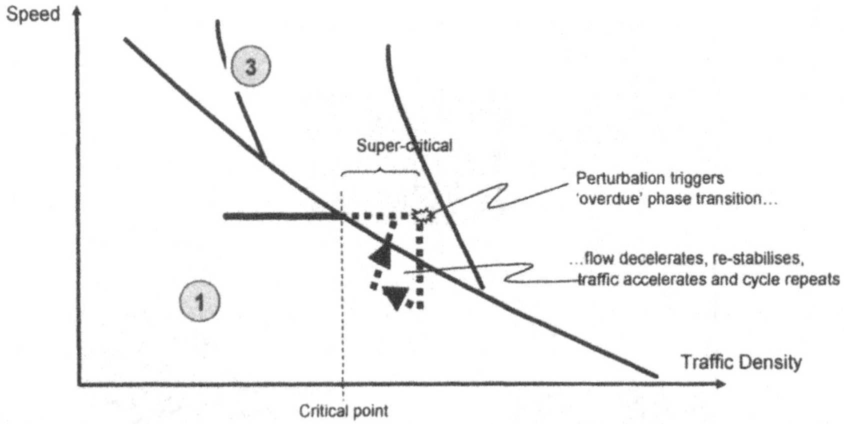


**Fig. 5.** Super-Critical Behaviour In Highway Traffic Flow And Influence Of Perturbations

The same thing occurs with traffic flow when any phase boundary is crossed. Figure 5 illustrates one such example. It is perfectly possible for the traffic flow to transition over a phase boundary without the transition occurring. However, like the initiation point in ice formation, as soon as an appropriate perturbation occurs in the traffic (a driver suddenly changes lane, new vehicles enter at an on-ramp, lanes narrow at road-works, etc) then the phase transition will occur. As shown in Figure 5, taking the transition between free-flow and wave-flow as its example, the perturbation causes the vehicle speed to drop suddenly. Once it drops sufficiently the phase boundary is crossed again and within a short time vehicles are able to accelerate, thus decreasing density. However the density decrease is temporary and so we get back to where we started in the cycle; flow at the original speed and density until another perturbation occurs.

### 2.2 Phase Transitions Define Contradictions

Returning to the idea that phase boundaries define contradictions, the connection we make here involves the idea that phase boundaries and contradictions both involve the concept of discontinuity. Just as the discontinuous shift between water and ice means presents a 'solid and liquid' contradiction, the same thing happens across the boundary between two traffic phases. Free-flow traffic behaviour is fundamentally different from pulse-flow behaviour. Just as ice and water can't be treated in the same way, we ought not to treat different traffic phases in the same way. In many ways, when TRIZ recommends the use of Principle 35, Parameter Changes it is a prompt for us to look not for optimizing parameter shifts, but rather shifts that transition across some kind of discontinuous boundary (Ref 6).

The idea that different phases act on different sides of a contradiction, is intended to give a clear indication that there is no such thing as an 'optimum' solution to this type of multi-phase problem. Thus any attempt to manage traffic using a single solution (by for example imposing a speed limit) is destined to work unsatisfactorily on both sides of the phase shift boundary. This phenomenon then connects us to another important idea, this time one from the world of cybernetics:

## 3 Law Of Requisite Variety

Cybernetics pioneer W Ross Ashby stated (Reference 7) 'only variety can absorb variety' – Figure 6. This apparently obvious and over-simple statement hides a mass of important ideas. As far as our traffic control problem is concerned, as well as re-enforcing the contradiction idea that there can be no single solution to a situation bearing multiple discontinuous phases, it informs tells us that there needs to be a level of variety in the potential solution that matches the variety present in the system we are trying to control. This means that the variety in the four-phases should by rights require different treatment.

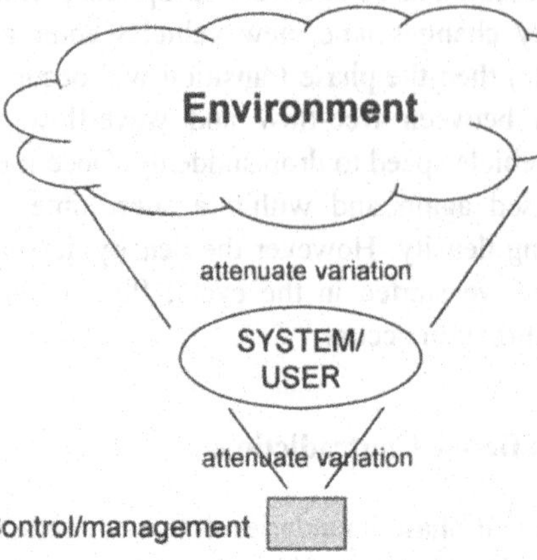


**Fig. 6.** Only Variety Can Absorb Variety

# 4 Solving The Contradictions

In geographical locations where the traffic problem has become bad enough, we can see how road planners have made at least one attempt to manage two of our four traffic phases. The basic contradiction for all four of the phases is one between speed and density (or 'speed' versus 'amount of substance' in TRIZ Contradiction Matrix terms). Figure 7 illustrates how this conflict pair has been mapped onto the latest version of the Matrix (Reference 7), and then how one of the suggested Inventive Principles in turn can be translated into the variable speed limit system now found in some busy road systems.

This 'variable speed limit' solution is an attempt to recognise the first mode ('free-flow') versus other modes (2, 3 and 4 in Figure 4) contradiction. The traffic control system in these solutions measures traffic density and sets the speed limit accordingly.

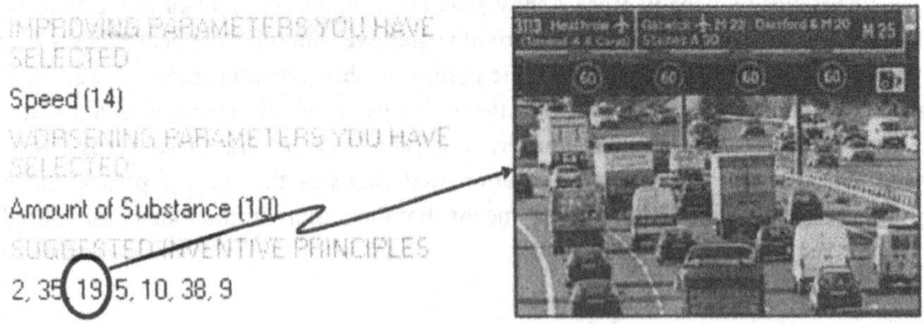


**Fig. 7.** Variable speed limit control systems work for one phase but not necessarily others

Variable speed limit systems represent a partial answer, but not (as the designers probably intended) the answer to the highway traffic congestion problem. The phase diagram and the Law of Requisite variety inform us why this is the case: Four discontinuously different phases require four different solutions. We experience the compromise presented by the current systems when we find ourselves sitting in a stationary car (i.e. mode 4) being told that the speed limit is 50mph. Choked/stop-start mode traffic requires a different solution than mode 3, 'pulsed flow' – where the variable speed limit solution actually works.

Knowing that four phases require four discontinuously different solutions causes us to return to the TRIZ Contradiction tools in order to identify further contradiction breaking solutions.

Staying with the contradiction parameters and Inventive Principle recommendations shown in Figure 7, the following conceptual solutions may help designers to better manage the four distinct traffic modes:

- Principle 5 (Merging) – combine the information generated at multiple road sensors in order to detect the presence or otherwise of 'pulsing' traffic behaviour. If no pulsing, then switch to an algorithm appropriate to other modes – e.g. in Mode 2 traffic, it may be advisable to display no message on the traffic signs.
- Principle 35 (Parameter Change) – in Mode 4 traffic it may be better to display another symbol other than a speed limit. For example a 'danger' or 'accident hazard' icon might be a better way to warn drivers that the traffic is in a dangerous state. A possible strategy here might be to display a delay time rather than a speed limit – playing on the well-known psychological phenomenon that people are far more likely to become stressed when they don't know what is happening and how long a problem will last than they would if they knew when the problem was likely to go away. Even if the news is bad (i.e. displaying a '15 minute delay' indication) it is preferable to no news because knowing the extent of the delay in this case would allow drivers to contemplate turning off their engine and reading a book.
- Principle 9 (Prior Counter-Action) – this suggests the idea of 'pre-tensioning' the traffic flow – perhaps by allowing a higher speed limit (i.e. above national limit) immediately after a blockage in order to unchoke the flow at the bottleneck. Or perhaps it might mean pre-warning drivers of emerging problems at longer distances further along the road – i.e. give drivers a message before they see the reduced speed limit signs.

The main point to note at this stage is that in this situation, only a computer simulation has been able to identify the full extent of the traffic problem. As far as traffic control designers are concerned, they have generated a partial solution to the problem (in all likelihood unconscious that they have solved a contradiction!), but partial is all it is. Computer aided innovation in this sense – and we think it generalizes to all complex, multi-phase problems – is about finding the contradictions that would otherwise be hidden. At the moment, the designer then needs to do the creative – solution generation – part of the innovation. But then this is the part most designers enjoy the most.

Later on, the optimizers can do their real world experiments and, with the help of the computers again (computers being excellent optimizers), design the actual quantified traffic control algorithms. This is, of course, a critical part of the solution implementation story, albeit it is one that has little if anything to do with the job of innovating.

## 5. Final Thought

The automotive industry already knows they can design automated vehicle systems – e.g. braking – that offer considerable safety benefits relative to even the best driver. And yet they also understand human psychology well enough to know that if control of critical activities like steering, accelerating or braking is taken away from the driver, the driver will not purchase the vehicle. People like to feel like they are in control of their own destiny. As suggested by the Philip Dick quotation at the beginning of this paper, CAI has a similar dilemma to the automotive designer. Take away control from the human designer by creating a 'computerised innovator' and, if the automotive analogy holds true, no-one will use it. On the other hand, present it in such a way that it offers new opportunities to employ their creative skills – as we think is the case with multi-agent programmable modelling environments – and there is a strong likelihood of a win-win situation; the computers do what they're good at (in this case finding what the human alone cannot); the human then gets to dream the breakthrough solution.

## References

1. Dick, P., 'Paycheck', Gollancz, 2004.
2. Naisbitt, J., 'Mind Set!: Reset Your Thinking and See the Future', HarperCollins, 2006.
3. Resnick, M., 'Turtles, Termites and Traffic Jams: Explorations in Massively Parallel Microworlds (Complex Adaptive Systems)', MIT Press, New Edition, 1997.
4. Grand, S., 'Creation: Life And How To Make It', Weidenfeld & Nicolson, 2000.
5. Wilensky, U., 'NetLogo', http://ccl.northwestern.edu/netlogo/. Center for Connected Learning and Computer-Based Modeling, Northwestern University, Evanston, IL, 1999.
6. Systematic Innovation E-Zine, 'Effective Use Of Principle 35', Issue 58, January 2007.
7. Ashby, W.R., 'An Introduction to Cybernetics', Chapman & Hall, London, 1957, available electronically at http://pespmc1.vub.ac.be/books/IntroCyb.pdf.
8. Matrix+ Software, www.systematic-innovation.com, 2007.

# Creative Tower Generated by Computational Intelligence

Xiyu Liu[1], Hong Liu[2]
[1]School of Management & Economics, Shandong Normal University, Jinan, China
[2]School of Information Science & Engineering, Shandong Normal University, China
Email: *xyliu@sdnu.edu.cn*

**Abstract.** Based on previous works by the authors on evolutionary architecture paradigm of evolving architectural form, new research is being carried out to develop a virtual environment with system implementation model of evolving creative tower designs. The solid models are created by evolution with computational intelligence and enhanced mathematical models. Models will evolve according to computational intelligence including generic algorithms, particle swarm optimization. Mathematical models include analytical functions, parametric functions and other nonlinear functions. We also present analysis of relationship between evolution and exploration.

## 1. Introduction

Computational intelligence and computer modeling have been efficient ways in architecture design (18). In this area much work has been achieved in natural model, evolutionary models and revolutionary models (Frazer, 2002). In fact, computer modeling has become so important that one can hardly find any design without the help of it. And there are more and more artworks and designs that are called generative art which have widely changed the conventional idea. As the work of professor Celestino Soddu shows (http://www.celestinosoddu.com/), generative art is one of the new ideas that can get artificial objects. In this way, we can work producing three-dimensional unique and non-repeatable shapes.

Of the many computer aided design tools up to now, most of them provide hand-made utilities, and a designer has to draw every sketch to get the outline of a product (Frazer, 2002). And many generative tools are based on sketches. These limitations motivate the conceptual design methodology introduced in this paper, which is capable of using mathematical functions to get novel shapes.

The purpose of this paper is to report our new research in evolutionary towers by computational intelligence. We will introduce some relations of computing to design

*Please use the following format when citing this chapter:*

Liu, X., Liu, H., 2007, in IFIP International Federation for Information Processing, Volume 250, Trends in Computer Aided Innovation, ed. León-Rovira, N., (Boston: Springer), pp. 33-43.

– a new bridge of design and enhanced computation algorithms. This is possible because there is usually not a single optimal design for any problem, but rather designs evolve. Concerning to computing design, we will introduce (1) computing tools that is useful for creating and improving design alternatives, (2) creation of conceptual resources that is helpful in order to create design concepts, (3) linear and nonlinear algorithm for improving basic design concepts towards successful solutions. A new system is developed to implement the design process with enhanced computation techniques and complex functions. The system kernel is compatible with object-oriented technology and component reusing. An evolutionary architecture paradigm with a focus on how visionary and creative forms can be achieved is demonstrated. We will present our complex form generation and visualization system with images and rapid prototyping models that are otherwise impossible to generate by normal CAAD systems without using generative and evolutionary computation.

A new type of genetic algorithm is studied for our generative design system. We extend the classical powerful techniques from modern nonlinear analysis theory to selection and optimization of GA. These techniques include topological spaces and partial ordering. A Zorn Lemma type of iterative procedure is introduced. This attempt will partially overcome the difficulty in implementing effective automatic selection in the application of genetic algorithms.

## 2. Mathematical models and 3D shapes

Apart from its computation functionality, mathematics indeed reveals the beauty of nature. From symmetry to structure, from honeycomb to skyscraper, even giving a glance to the crowding cars, we will find the beauty of mathematics everywhere.

However, it is not easy to look into this beauty without the help of artist or computer. Of the early literatures, the book of Gerd Fischer (Gerd, 1986) is a successful one that introduces the elegance and beauty of mathematics. This book collected 132 images taken from real models of the most important mathematical models, including differential geometry, projective models. It is this book that changes the viewpoint of many people who always take mathematics as abstract, and who reveals the beautiful and symmetric structures that are potentially virtual models of architectures.

Nowadays, one of the most significant ways to understand a mathematical model is computer visualization. However, due to the fully nonlinear nature of many functions, it is not an easy work to develop accurate shapes for general nonlinear functions. One way to solve this problem is the finite element method. There are several approximate techniques for nonlinear functions. The simplest is planar piece. In his work, Peter J. Bentley (1996) uses primitive shapes consist of a rectangular block or cuboids with variable width, height and depth, and variable three dimensional positions to construct nonlinear objects. His blocks are intersected by a plane of variable orientation in order to approximate curved surfaces. More accurate techniques include nurbs approximation, polynomials approximation and others. It should be noted that one can hardly find the balance between a better approximation and acceptable computing time.

With the help of solid modelling libraries, we developed 3D solid model visualizations for complex functions in this project. A prototype system has been implemented based on an integration of ACIS 3D solid modelling kernel and MatLab with a C++ graphical user interface. Our basic geometrical objects for approximation are nurbs surfaced units. Our system is fully compatible with commercial CAAD tools and systems, as well as rapid prototype facilities. A large number of object-oriented components of sophisticated surfaces and envelopes based on taxonomy of generic form have been built. In particular, complex forms are classified as linear, quadratic, trigonometric function, exponential functions, root functions compounded functions, rotations, sphere and cylinder co-ordinates, implicit function. Computational mechanisms have also been developed with which these basic data structures and components can be visualized, combined or split to allow new data structures or new forms to be derived using generative techniques.

## 3. Theoretical structure: the concept of homotopy

What is homotopy? Intuitively, a homotopy is a constant deformation from one shape to another according to some rules. Although homotopy itself is an important concept in geometry and topology, we only borrow the idea here to describe our problem. These rules are called homotopy map. By defining various maps we can generate different homotopies. However, we should remember that there are indeed shapes that are not homotopic, that is, we can not find homotopy to transform one into another.

Fig3.1 is a curve illustrating homotopy. The left one is the original curve while transforming to the final curve, the right one. Intermediate curves show that homotopy is a continuous shaping process.

Whenever we use homotoopy to generate towers, the outline of the building is often the most attracting factor. This is the second factor just following its high and extruding illusion in a city landscape. In fact, a special tower in a city even becomes the representation of the city. In this paper, our main concern is to study generated towers with outlines determined by mathematical functions and evolution. We use a multi-coding schema to represent phenotypes, that is, the continuous schema corresponding to continuous functions, and the discrete schema corresponding to discretization of functions.

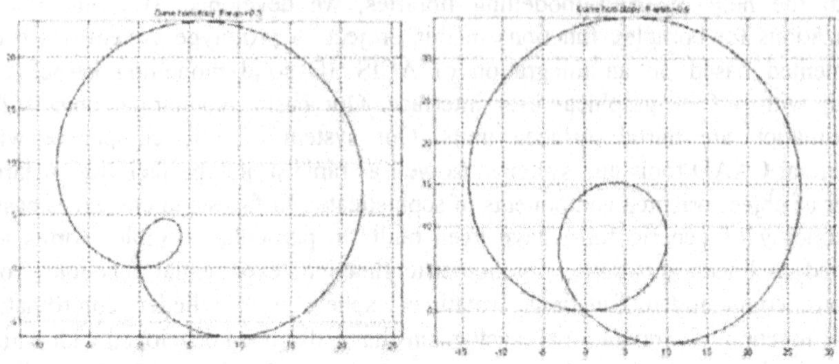

**Fig 1.** The two images are Homotopic curves (parameter = 0.3 and 0.6 respectively).

It is well known that one of the most important and difficult problem in evolution based applications is the appropriate definition of fitness function. It is this function that determined the selection and optimization of the evolution and the final solution. In the literature, many authors use the artificial selection technique which indeed can solve part of this problem, but delimitates some of the evolutionary nature of the problem and adds some uncertainty caused by human determining.

To solve this problem and for the use of the specific application in this paper, we apply a multi-objective fitness function for optimization. And another feature is that we use a partial ordering technique derived from nonlinear analysis to represent the complicated relationship in candidate solutions from the population. Fig3.2 is a diagram of the functional units of the system.

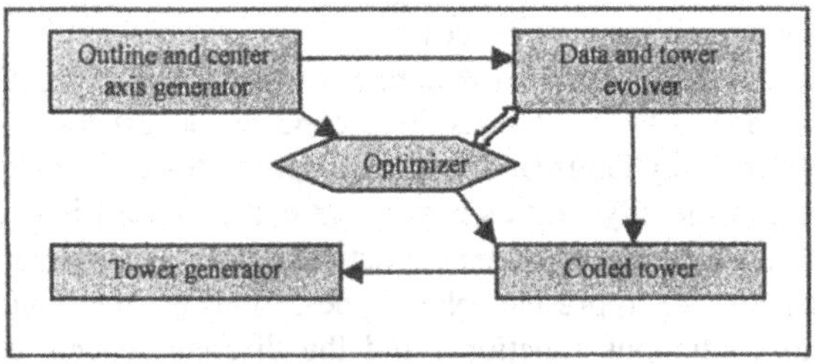


**Fig 2.** System functional units

## 4. Genetic algorithms and evolutionary models

Having become widely used for a broad range of optimization problems in the last ten years, the GA has been described as being a "search algorithm with some of the innovative flair of human search". GAs are today renowned for their ability to tackle a huge variety of optimization problems (including discontinuous functions), for their consistent ability to provide excellent results and for their robustness. Natural evolution acts through large populations of creatures which reproduce to generate new offspring that inherit some features of their parents (because of random crossover in the inherited chromosomes) and have some entirely new features (because of random mutation). Natural selection (the weakest creatures die, or at least do not reproduce as successfully as the stronger creatures) ensures that, on average, more successful creatures are produced each generation than less successful ones. As described previously, evolution has produced some astonishingly varied, yet highly successful forms of life. These organisms can be thought of as good 'solutions' to the problem of life. In other words, evolution optimizes creatures for the problem of life.

Any evolutionary models require architectural concepts to be described in forms of genetic codes. Then these codes are mutated and developed by computer programs into a series of models called population. While models are evaluated in optimization or selection sub-systems, the codes of successful models are constantly picked up until a particular stage of development is selected for prototyping in the real world.

In order to achieve the evolutionary model it is necessary to define the following: a genetic code script, rules for the development of the code, mapping of the code to a virtual model and, most importantly, the criteria for selection. The representation of phenotypes is a fundamental element of any evolutionary system. Phenotypes will define the representation of designs, which will formulate all allowable solutions that can be evolved by the system. Moreover, the phenotype representation plays a significant role in determining the size and complexity of the genotype. The two main representation methods are the surface representations (or boundary representations) and constructive solid geometry (CSG). The first method typically uses combinations of equations and control points to specify shapes, while CSG combines different primitive shapes to form more complex shapes. There are a third of the commonly used solid representations which are called spatial partitioning. This is to decompose a solid into a collection of smaller, adjoining, non-intersecting solids that are more primitive than the original solid. There are a number of variations including: cell decomposition, spatial-occupancy enumeration, octrees, and binary space-partitioning trees.

Primarily, our model in this paper is the boundary representation model. In this representation, towers will be represented by the mathematical data of the main model geometry. Changing the surfaces would be simply achieved by adding or subtracting the mathematical data. The second problem is the data model. In accordance with the mathematical functions type data, we will use the cell division model. A cell division model is based on the structure of a living creature. As in nature, the shape of a living creature is constructed from the basic genetic information to the cells and organisms. The genotype contains information that is the basic construction unit of everything, called the chromosome. Chromosomes will form proteins and other large molecules. Chains of molecules will form tissues and organisms till the whole body. In natural environment, development begins with the chromosome, which forms the base. Then a number of smaller cells are constructed. Large cells are resulted from joining and other operations and form a multi-cellular structure. In a word, the cell division models simple divide the whole into a basic units and operations.

For better description of the combinations of functions, we will use the item jelly model. This is a derivation of the cell model. A layer called the jelly layer is added to represent a compounded structure. We will use this model to represent functions in various combinations. Basically, the model has three layers, that is, the gene layer, the cell layer and the jelly layer.

The second model in our project is the discrete model. The two basic structures in this model are the section and outline data structure. Each of these two outlines is an ordered set of double numbers, with auxiliary data indicating steps and size. To eliminate the discontinuities caused by data, a mollifier operation is introduced.

Mollifiers act as a function smoother. They can smooth a discontinuous function by substituting the value at one point by some average in a neighbourhood. In the one dimensional case, a mollifier is a nonnegative, real-valued function $J \in C_0^\infty(R)$ such that $J(x) = 0$ if $|x| \geq 1$, and $\int_R J(x)dx = 1$.

## 5. Particle swarm optimization

Particle swam optimization is a stochastic optimization technique motivated by the behavior of a flock of birds or the sociological behavior of a group of people. There are many improvements to the original PSO. In this section, we briefly introduce the traditional PSO and its variations.

### 5.1 Basic PSO algorithm

The PSO is a population based optimization technique, where the population is called a swarm. Each particle represents a possible solution to the optimization problem. Unlike evolution process, PSO does not use genetic operations. Instead, particles fly in the n-dimensional search space according to a speed.

Suppose that $x_i = (x_i^1, x_i^2, \cdots, x_i^n)$ is the current position of the particle with index $i$. We use $v_i = (v_i^1, v_i^2, \cdots, v_i^n)$ to represent the current velocity of particle number $i$. Let $p_i = (p_i^1, p_i^2, \cdots, p_i^n)$ be the current personal best position of particle $i$ and $f(x)$ be the target function which will be minimized. Now we write the size of the population, that is, the number of all particles, by $s$, and denote the current global best position found by all particles during previous steps be $p_g(t)$. Therefore, the evolution equation of basic PSO is

$$\begin{cases} v_i(t+1) = \omega v_i(t) + c_1 r_1(t)(p_i(t) - x_i(t)) + c_2 r_2(p_g(t) - x_i(t)) \\ x_i(t+1) = x_i(t) + v_i(t+1) \end{cases} \tag{5.1}$$

Where $c_1$ and $c_2$ are the acceleration coefficients, $r_1$, $r_2$ are elements from two uniform random sequences in the range (0,1).

### 5.2 PSO with constriction factor

An alternative version of PSO incorporates a parameter called the constriction factor and the swarm is manipulated according to the equations 12.

$$\begin{cases} v_i(t+1) = \chi(v_i(t) + c_1 r_1(t)(p_i(t) - x_i(t)) + c_2 r_2(p_g(t) - x_i(t))) \\ x_i(t+1) = x_i(t) + v_i(t+1) \end{cases} \tag{5.2}$$

Here $\chi$ is the constriction factor, and denote the cognitive and social parameters respectively. It is usually chosen with uniform distribution in the interval [0,1]. The value of the constriction factor $\chi$ is typically obtained through the formula $\chi = 2\kappa / \left|2 - \phi - \sqrt{\phi^2 - 4\phi}\right|$ where $\phi > 4$ , $\kappa = 1$ , and $\phi = c_1 + c_2$ . Different configurations of $\chi$ , as well as a theoretical analysis of the derivation of $\chi$ can be found in 12.

## 5.3 LBest model

Since the original PSO uses global best position, it is called Gbest model. In order to avoid premature convergence, R. C. Eberhart, P. Simpson, and R. Dobbins 13 proposed the Lbest model. Instead of using a unique attractor, the Lbest model uses multiple attractors. This approach divides the population into multiple neighborhoods where each neighborhood maintains a local best position.

The evolution equation is described as in (5.3). Notice that the neighborhood $N_i$ does not related to the actual position of particles in the searching space. Instead, it is only related to the coding, or index position of particles.

$$\begin{cases} N_i = \{p_{i-l+\sigma} \mid \sigma = 0,1,\cdots,2l\},\ x_i(t+1) = x_i(t) + v_i(t+1) \\ p_i(t+1) \in \{N_i \mid f(p_i(t+1)) = \min_{x \in N_i} f(x)\} \\ v_i(t+1) = \omega v_i(t) + c_1 r_1(t)(p_i(t) - x_i(t)) + c_2 r_2(p_g(t) - x_i(t)) \end{cases} \tag{5.3}$$

## 5.4 Improved PSO

There are many literatures focusing on variations and improvements of the traditional PSO. For the purpose of this paper, we present several aspects of these researches. Firstly in 1998, P.J. Angeline in 14 proposed a generalized PSO based on selection, the tournament selection method. In this paper, an individual is compared with others by fitness value. Next sort the population and replace part of the population with worst fitness by other individuals with better fitness. The procedure of the algorithm is as follows.

(1) Select an individual and compare its fitness value with every other individual. If it is better, then add one score to itself. Repeat this procedure until every individual has a score.

(2) Sort the population according to its score.

(3) Copy half of the population with higher score and replace the other half with lower score.

In another paper 15, P.J. Angeline added reproduction to the traditional PSO. According to a user predefined reproduction probability, some individuals are selected into a crossover pool. Particles in this pool crossover pair-wisely and reproduce the same number of offsprings. Then parents are replaced by offsprings.

The whole number of the population remains changed. The crossover operator is shown as follows.

$$\begin{cases} x_a(t+1) = r_1 x_a(t) + (1-r_1) x_b(t) \\ x_b(t+1) = r_1 x_b(t) + (1-r_1) x_a(t) \end{cases}, \begin{cases} v_a(t+1) = [v_a(t)+v_b(t)] \|v_a(t)\| / \|v_a(t)+v_b(t)\| \\ v_b(t+1) = [v_a(t)+v_b(t)] \|v_b(t)\| / \|v_a(t)+v_b(t)\| \end{cases} \quad (5.4)$$

### 5.5 Niche technology and PSO

There are already several authors taking efforts to combine niche technology with PSO. One of these efforts is made by P.N. Suganthan 16. This is a variation of the Lbest model but neighbor is defined by space position rather than index position. In each generation, distance between particles is calculated. The maximum distance is marked by $d_{max}$. For ever pair of particles, the ratio $\|x_a - x_b\| / d_{max}$ is taken as measure to define the neighbor, where $\|x_a - x_b\|$ denotes the distance between them. The neighbor of each particle is dynamic in the sense that this neighbor grows from a one particle neighbor (the particle itself) to the whole population finally.

In another effort, J. Kennedy 17 discussed effects of neighbourhood topology to the performance of PSO. He proposed several basic neighbourhood topologies: the ring topology, wheel topology and its generalizations.

The most promising technique will be the integration of genetic algorithms with PSO. This technique is achieved with population, or subpopulations, which are evolving and flying simultaneously. In order to keep the genetic operations and the flying operations together, we have two models, i.e., the Genetic PSO, and the Swarm Evolution. Fig 5.1 shows the basic architecture of the two paradigms.

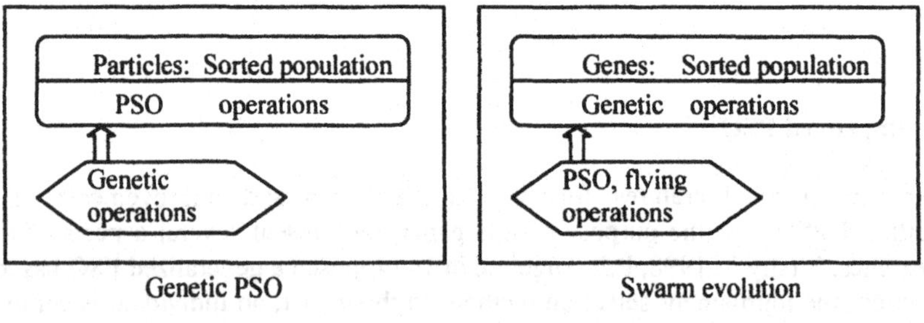


**Fig 3.** The Genetic PSO and swarm Evolution paradigms.

## 6. System implementation and generated towers and architectures

Based on the theory described above, we build system implementations based on solid modeling techniques. Two integrated design studio environments have been developed under this project. The platform is personal computers under windows 2000 or windows XP of Microsoft. The programming languages are Microsoft

Visual C++ version 6.0, Acis 3d Kernel, and MatLab version 6.1 of Mathworks. These systems are implemented based on an integration of ACIS 3D solid modeling kernel and MatLab with a C++ graphical user interface. One of the features of the systems is that they are fully compatible with commercial CAAD tools and systems, as well as rapid prototype facilities. A large number of object-oriented components of sophisticated surfaces and envelopes based on a taxonomy of generic form have been built using evolutionary techniques and partial ordering theory, Computational mechanisms have also been developed with which these basic data structures and components can be visualized, combined or split to allow new data structures or new forms to be derived using generative techniques. We are also exploring the possibility to scale up the applications with potentially thousands of solid objects with textual and spatial design details in our Global Virtual Design Studio powered by high performance computer and multiple VR projection facilities.

Now we describe some specific features of the design studio. The first one is named as TowerDev. Features include fully controllable solid modelling environment; nonlinear transform of existing solid models to give new designs; Acis sat viewer; fly through viewer; colour and rendering; programmable design for architecture. The main user interface is as in Fig 6.1.

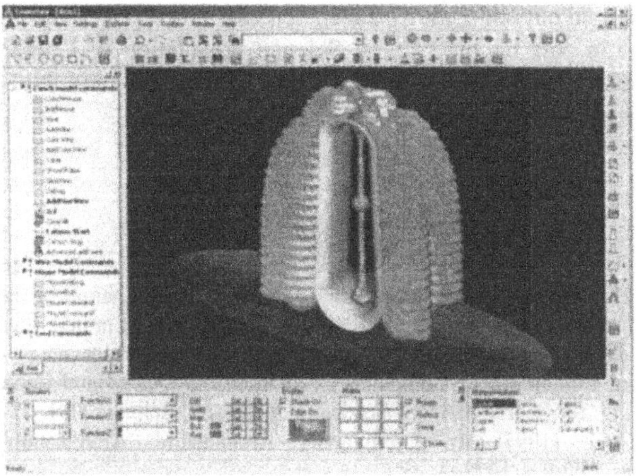

**Fig 4.** Generated house by TowerDev.

Now we give some more examples of generated towers and other solid models by the studio TowerDev.

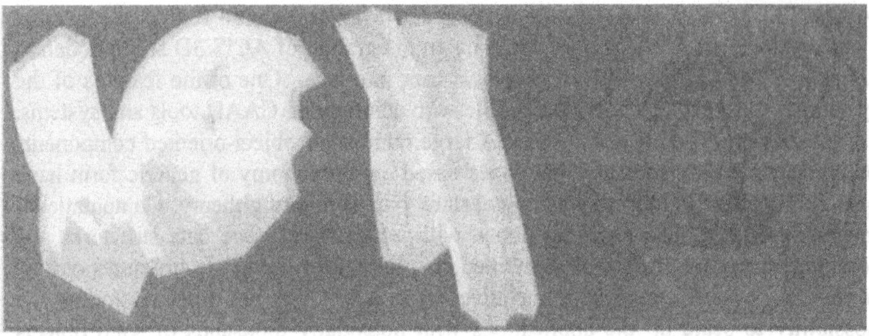

**Fig 5.** Generated towers with rendering.

## 7. Acknowledgements

This work is carried out under the "Taishan Scholar" project of Shandong China. It is also supported in part by the Natural Science Foundation of China, the Natural Science Foundation of Shandong Province (No. Z2004G02), and the Scientific and Technology Project of Shandong Education Bureau J05G01.

## References

1. Frazer, J. H. (2002) A natural model for architecture —The nature of the evolutionary model. In Cyber Reader, edited by Neil Spiller. Phaidon Press Limited. London, 2002.
2. Frazer, J. H. (2001) Design Workstation on the Future. Proceedings of the Fourth International Conference of Computer-Aided Industrial Design and Conceptual Design (CAID & CD '2001), International Academic Publishers, Beijing, 2001; 17-23.
3. Frazer, J. H. (1995) An Evolutionary Architecture. Architectural Association Publications, London, 1995.
4. Frazer, J. H. (2000) Creative Design and the Generative Evolutionary Paradigm. In P. Bentley ed. Creativity and Design. In press. 2000.
5. Gerd Fischer (editor), Mathematical Models, Friedr. Vieweg & Sohn Verlagsgesellschaft mbH, Braunschweig, 1986.
6. Tang, M. X. Knowledge-based design support and inductive learning. PhD Thesis, Department of Artificial Intelligence, University of Edinburgh, 1996.
7. Tang, M.X. A knowledge-based architecture for intelligent design support. The Knowledge Engineering Review, 1997;12(4):387-406.
8. Peter J. Bentley (1996), Generic evolutionary design of solid objects using a genetic algorithm. Thesis of doctor of philosophy, University of Huddersfield.
9. Xiyu Liu, Mingxi Tang and John H. Frazer (2002) Shape reconstruction by genetic algorithms and artificial neural networks. Proceedings of The 6th world Multiconference on systemics, cybernetics and informatics (SCI2002).
10. Foley, J., van Dam, A., Feiner, S., Hughes, J. (1990). Computer Graphics Principles and Practice (second edition). Addison-Wesley.
11. Goldberg, D. E., (1989). Genetic Algorithms in Search, Optimization & Machine Learning. Addison-Wesley.

12. M. Clerc and J. Kennedy, "The particle swarm-explosion, stability, and convergence in a multidimensional complex space," *IEEE Trans. Evolutionary Computation*, vol. 6, pp. 58–73, Feb. 2002.
13. R. C. Eberhart, P. Simpson, and R. Dobbins, Computational Intelligence PC Tools: Academic, 1996, ch. 6, pp. 212–226.
14. P.J. Angeline, "Evolutionary optimization versus particle swarm optimization: philosophy and performance differences", in V. William Porto, N. Saravanan, Donald E. Waagen, A. E. Eiben (Eds.): *Evolutionary Programming VII, 7th International Conference, EP98*, San Diego, CA, USA, 1998, pp. 601-610.
15. P.J. Angeline, "Using selection to improve particle swarm optimization", in *Proc. IEEE World Congress on computational intelligence, ICEC-98*, Anchorange, Alaska, 1998, pp.84--89.
16. P. Suganthan, "Particle swarm optimiser with neighbourhood operator", in: Angeline, P. J., Michalewicz, Z., Schoenauer, M., Yao, X., Zalzala, A. (eds.): *Proceedings of the Congress of Evolutionary Computation*, Vol. 3. IEEE Press, 1999, pp.1958-1962.
17. J. Kennedy, "Small worlds and mega-minds: effects of neighborhood topology on particle swarm performance", in *Proc. Congress on Evolutionary Computation*, Piscataway, NJ: IEEE Service Center, 1999, pp.1931–1938.
18. R. Kicinger, T. Arciszewski, and KD Jong, "Evolutionary computation and structural design: A survey of the state-of-the-art", Computers & Structures, Vol. 83, No. 23-24, pp. 1943-1978, September 2005.

# General Theory of Innovation

## *An Overview*

Greg Yezersky
Institute of Professional Innovators (IPI); 35987 Chater Crest Road, Farmington Hills MI 48335 USA; Website: www.ipinetwork.com; Email:gyezersky@ipinetwork.com

**Abstract.** The odds for success of a future CAI system (as well as the present CAI movement) are completely dependent on the quality of an underlying theory of innovation and the effectiveness of its tools, processes and models. This paper establishes a set of requirements for such a theory, evaluates existing approaches, methodologies and theories (including TRIZ), and presents an overview of the General Theory of Innovation (GTI) that, in the author's opinion, satisfies most of the established criteria. The overview includes the theoretical foundation of GTI, a list of available applications, a list of future tasks, and other pertinent information.

**Key words:** General Theory of Innovation; evolution; natural laws; value; market.

## 1 Introduction

The effectiveness of any computer-based process largely depends on the quality of models that are used for the design of a respective software package. The models quality, in turn, is absolutely predetermined by a degree of sophistication of our understanding of the piece of reality that we try to computerize. The field of innovation and the recently emerged attempts to computerize innovation-related activities (by the global CAI community) are no exception from the above rule. Acceptance of this position leads us to the need to objectively evaluate existing innovation theories, methodologies and techniques, identify issues (if any) that need to be addressed, and solve them. Objective evaluation, in turn, requires defining innovation and establishing a set of judgment criteria.

---

*Please use the following format when citing this chapter:*

Yezersky, G., 2007, in IFIP International Federation for Information Processing, Volume 250, Trends in Computer Aided Innovation, ed. León-Rovira, N., (Boston: Springer), pp. 45-55.

### 1.1 Defining innovation

First of all, we have to recognize that there are many different definitions [1] of what innovation is. Also, the global community, including our colleagues, believes that there are many different kinds of innovation. J. Schumpeter [2], for example, distinguished between five different types of innovation: new products, new methods of production, new sources of supply, the exploration of new market, and new ways to organize business. Any theory and any definition of innovation need either to embrace this complexity or find a plausible way to simplify the situation.

The goal of simplification can be achieved because all these objects of innovation (product, process, service, as well as various entities from an organization to a country) are systems. If it is possible to create a theory of innovation for systems, in general, there would be no significant difference between all these types of innovation that are often mentioned in books; they would be different applications of the same theory. Another opportunity to find a common denominator comes from the fact that each innovation (regardless of whether it is a product, process, method of marketing, business method, market, etc.) is just a visible result of a respective process (Fig. 1) that is hidden and thus is not mentioned. The point is that it is the process with its focus on a change that determines the result. It is also worth noticing that the content of an innovation, for which it is judged by the market, is defined by the end of the conceptual design stage. As a result, the author defines innovation as follows.

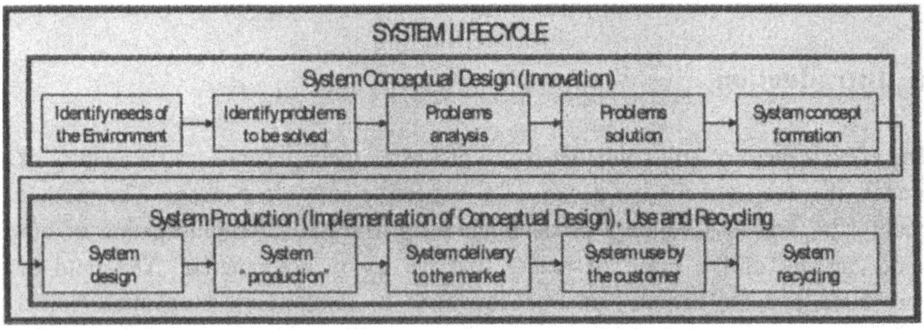


**Fig. 1**. The Lifecycle of a System

***Definition 1.***
*Innovation is a process of value creation, which consists in changing the composition of a set of variables describing a system.*
***Definition 2.***
*Innovation is an outcome out of the process that fits the definition 1.*

While the second definition enables alignment with a "typical" understanding of what innovation is, the first (the primary!) definition provides most of the benefits.

- First, the definition breaks down the process of innovation into a rigid set of stages, each having its own unique goal, input and output. Thus, the further work can (and should) go in the direction of defining them, identifying the most effective tools, processes and best practices for each of the stages.
- Acceptance of the innovation as a process clearly points at the need to control each separate stage of this process in order to avoid inconsistency (variability) of results, as it is prescribed by the operation management theory and various quality methodologies.
- The corporate inability to control the process of innovation explains the phenomenon of inevitable growth deceleration, stagnation, loss of market share and eventual fall of the market leaders, which was confirmed by professors Kim and Mauborgne [3]. Since today organization don't know how to control the stage of innovation, the results of competition are random, and sooner or later the sheer force of probability determines the fall of present leaders (consider recent examples of Dell and GM).

### 1.2 A set of requirements for evaluating a theory of innovation

One of the benefits of defining innovation as a process is the possibility to define a set of requirements that any theory of innovation must satisfy. Here they are.

1. A theory must have the capability to address identified issues; analyze and solve existing problems;
2. A theory must have predictive capabilities and identify future needs (future problems) of a respective system's "Environment";
3. A theory must provide objective criteria for judging novel concepts; especially, the theory MUST provide means to evaluate the upcoming innovations potential for their future success or failure in the marketplace;
4. A theory must be objective; maximally independent from its user;
5. A theory must be universal; work for a system of any nature.

### 1.3 Evaluating existing theories, methodologies and techniques

Consideration of multiple theories [4–13], both past and contemporary, shows that all of them fail to satisfy the totality of the established requirements. Even TRIZ [14, 15], which is the most advanced theory, also does not meet the set criteria.

1. TRIZ, by being based on technology, is not a universal theory, which makes it inapplicable to those industries and firms that are not technology-based (e.g. banking, retail, etc.) without adapting its tools first.

2. Even technology-based companies have innovation needs for activities that are not technology-based; for instance, HR, P&S, strategy, etc.
3. Finally, while being able to improve products and processes, TRIZ cannot answer the most important question of all; namely, what are the odds of market success that a new concept has? The reason for this inability is again in the foundation of TRIZ that focuses on technical systems while the market that ultimately determines future success or failure of an innovation is NOT a technical system.

This situation naturally creates three distinct possibilities for creating an innovation theory that would satisfy all (or the majority) of the above established criteria.

1. Further evolve TRIZ as the theory that presently meets most of the requirements trying to expand its boundaries beyond the present limits.
2. Combine TRIZ with other existing theories, methodologies and techniques (such as marketing; QFD, Axiomatic Design, AHP, etc.) to produce desired results.
3. Create a new theory from scratch so that it would overcome shortcomings of its predecessors and satisfy all or most of the requirements.

## 2 General Theory of Innovation (GTI)

Guided by the above requirements, in 1987 the author of this paper deliberately chose the third option, which ultimately resulted in the creation of the General Theory of Innovation (GTI). Right from the start, three crucial choices were made. Just as TRIZ, the process of creating GTI was based on the historical analysis of evolutionary processes of systems. Contrary to TRIZ, the systems were deliberately chosen of different nature. Last but not least, the focus of the investigation was not on the systems themselves but on the relationships they had with their respective Environments. This means that not only solutions were analyzed, but also the problems that caused the need for these solutions as well as the conditions that made these solutions successful. Here are a few examples of the systems that were investigated.

- Sound storage medium has evolved from Edison's phonograph, to wax cylinders, to discs with lateral grooves, to double-sided discs, to reel-to-reel magnetic tapes, to 4- and 8-track tape cartridges, to compact cassettes, to CD, to DVD, to MP3.[16]
- The use of currency evolved from the barter of goods (cattle, grain, etc.), to silver ingots guaranteed by Cappadocian rulers (2200 BC), to the first crude coins made from naturally occurring amalgam of gold and silver (640 BC), to Chinese paper money (800 AD), to bank-backed notes (1633 – 1660), to the first credit card (1950s), to electronic money.[17]
- Message delivery evolved from sending a messenger on foot, to a messenger on horseback, to the creation of regular mail service, to mail

service supported by cars, trains and planes, to fax, to the next day service, to e-mail.

Despite being very different, all three examples have a number of things in common.

- ***Any product or service (process) is a system.***
  This means that each and every product or service represents the union of parts or procedures connected to each other in order to deliver value to the customers. No individual element of a system can deliver the same value on its own.
- ***Systems (products, services, industries) evolve.***
  Systems evolve over time to adapt changes in customers' needs and desires.
- ***Systems evolve in the predominant direction.***
  The course of a system's evolution coincides with the delivery of ever-increasing performance while requiring less resources for providing that performance

The predominant direction of evolution can be expressed as the ratio of the sum of the functions of a system (an embodiment of performance) to the sum of connections the system needs to establish for obtaining the required resources for achieving the functionality. While functioning is easily understood, let's discuss connections in greater details. The first connection to be considered is the "customers expenditures" list (effort needed for use, time involvement, cost of ownership, space for storage, etc.), followed by requirements such as materials, energy, number of manufacturing processes and suppliers, production time, as well as sub-categories and consequences such as scrap, wastes, pollution, etc. Through the relationship between function and connection, this ratio, entitled the ***Coefficient of Freedom*** (any function empowers a system and makes it freer while any connection increases its dependency and decrease freedom), embodies the business world concept of value. The greater the Coefficient, the greater the value delivered by a product or a service.

$$C_{Freedom} = \frac{\Sigma\ Functions}{\Sigma\ Connections}$$

Historical analyses of the evolutionary process for various systems (those above, as well as bicycles, glass making, baking equipment, welding, shopping, banking, car, movie renting, publishing, the computer mouse, the car door hinge, safety airbags, etc.) clearly show the validity of the Coefficient of Freedom. It is universal, whether it is applied to products, processes, services, or various entities such as organizations (both for profit and not-for-profit), industries, markets, regions, etc. Moreover, these analyses lead firmly to the conclusion that ***systems do not evolve randomly; the evolutionary cycle of all systems, regardless of their specific nature, is governed by the same set of natural laws that are completely independent of human will and desire***, which is the major postulate of the General Theory of Innovation (GTI), first defined in 1988. The natural law governing the process of evolution (growth, expansion) of various systems states that "***the direction of a system's evolution coincides with a continuously increasing degree of freedom of this system's Environment***" and is thus entitled the Law of an Increasing Degree of Freedom.

## 3 Major Implications / Corollaries

Acceptance of the GTI foundation, which is existence of the Natural Laws governing the process of evolution, automatically leads to the following benefits and gains that are direct corollaries (natural consequences) of the accepted position.

### 3.1 The nature of a challenge (problem, failure)

The nature of any challenge/problem/failure experienced by a system is in a deviation from the direction prescribed by the Natural laws of evolution. Consider an analogy of disobeying the natural laws of traffic on a freeway (driving against the traffic, changing lanes continually, driving with a speed that significantly differs from the one of the flow, etc.), which always elevates the risk and creates problems. Being able to efficiently identify the origins of problems, which are always a result of our choices, greatly improves our abilities to effectively address them by going to the root cause and restoring a "lawful" behavior.

### 3.2 The nature of success

On another hand, the nature of success is in the obeying the "LAWS". There is no exception from the rule. Just as we must follow the laws of physical science when designing products or services if we expect these products or services to work well, we must also follow the laws of evolution if we expect business success. Today's executives, whether they know it or not, follow these laws when they succeed. However, they do so intuitively but not consistently or methodically, thus producing very mixed results. GTI articulates evolutionary laws and introduces a set of tools for working consciously and strategically within the laws.

### 3.3 The capability to forecast the future of evolution

Knowledge of a system location on the evolutionary curve combined with knowledge of the evolutionary Laws allows any organization to forecast the system (product, process, service, etc.) future with a great degree of precision.

### 3.4 The capability to objectively judge upcoming innovations

Existence of natural Laws of the evolutionary cycle has enabled creation of the objective criteria for evaluating proposed innovations, the importance of such criteria being self-evident. At the time of working on a direct-current motor, Thomas Edison, completely dismissed the efforts by George Westinghouse stating that alternating current was nonsense, which had no future. Every innovation improves a system, moving it along the evolutionary curve. Whether this move complies with the laws (or deviates from the laws) constitutes a criterion for evaluating the innovation.

### 3.5 The capability to control the process of innovation

With above capabilities, one can control the entire process of innovation (as defined above) thereby greatly reducing risk and variability of results, increasing manageability of the process and ROI of R&D. Finally, while understanding that GTI (just as any other scientific theory) can be endlessly perfected, it, in principle, meets the criteria set at the beginning of this paper.

## 4 Available Applications and Tools

After 20 years, the following applications and tools have been developed based on the knowledge of the natural laws of evolutionary cycle.

### 4.1 Analysis and solution of complex systems-related problems

As we discussed previously, the essence of any problem is the fundamental conflict between the choices we made while pursuing our goals and the Natural Laws of evolution. The process essence is in identifying these choices that led to the conflict and correcting them. To accomplish these goals, the following tools were created up-to-date: RelEvent™ Diagram; Problem/Solution Templates™; the Algorithm for Conflict Elimination (ACE); Generic strategies for conflict elimination; and so on.

### 4.2 Carrying out complex projects

When addressing a system-related problem, it is assumed that the nature of dissatisfaction is associated with a very specific aspect of the system performance: noise, strength, etc. Complex projects, as GTI defines them, relate to such important for every organization activities as cost reduction, quality, reliability, performance and productivity improvement as well as failure prevention. The reason for being called complex is that any of the above activities can be reduced to identification of those multiple (hence complex) problems, presence of which causes emergence of high cost (or low quality, reliability, etc.), and subsequent solution of the identified problems. All the tools, techniques and principles, which were used for analysis and solution of a single or stand-alone problem, will be also effective and valid for efficient achievement of the goals of a complex project.

## 4.3 Forecasting the future of the evolution of a system

Knowledge of a system location on the evolutionary curve combined with knowledge of the evolutionary Laws allows any organization to forecast the system (product or service) future with great degree of precision. The entire procedure of forecasting the future of a system consists of two major stages. First, by using the Laws future problems, which will cross the path of your system, are identified, and then they are solved by using the problem solving tools previously discussed.

## 4.4 Innovation assessment and tools for decision-making

Existence of natural Laws of the evolutionary cycle has enabled creation of the objective criteria for evaluating proposed innovations, the importance of such criteria being self-evident. Compliance with the evolutionary laws (or deviation from the laws) constitutes the foundation for evaluating an innovation.

## 4.5 Patent circumvention or patent protection against circumvention

At the heart of any patent, there is a solution for a problem. Patent circumvention then is finding an alternative solution for the same problem; or finding and solving an alternative problem for the same goal; or finding an alternative goal, followed by identification of a problem needed to be solved to reach the goal and subsequent solution of this problem, for which tools are available. The patent protection against circumvention is the opposite procedure and is carried out in the similar fashion.

## 4.6 Strategic management (business applications)

GTI states that innovation in the area of strategic management (identification of a change required for repositioning an organization with the purpose of obtaining competitive advantage) is immeasurably more important than innovation in any other area of corporate activities such as product or process innovation. The reasoning behind this very firm position is simple: the history of business definitely shows that companies with inferior products but superior strategies beat their technically superior competitors. Examples abound: Microsoft vs. Apple; Dell vs. IBM and Compaq; Big 3 vs. Tucker Corporation (founded by Preston Tucker).

Knowledge of the evolutionary laws is applicable not only to such systems as technology-based products, services, processes but also to the organizations (both for-profit and not-for-profit), industries and markets, which are also the systems. Moreover, application of GTI to the strategic management enabled creation of specialized tools such as Generic Growth Strategies; Value Matrix; Value Growth Templates and other. If an organization can precisely forecast the future of its own products and processes as well as foresee where the market will go, this company can use this knowledge at any moment for creating new powerful strategies, finding new markets for products and services, finding new sources of revenue, generating and controlling growth. This company will have a substantial advantage comparing to its uninformed rivals, which is the solid foundation for continuous advantage and success.

### 4.7 Strategic Innovation

Not all innovations are born equal! Out of the minority that are financially successful, a very few are capable of moving the markets and increase the market share for their creators. The deliberate (on-demand) creation of these innovations is the essence of this application that involves analysis of such systems as the market, a respective company with the focus on its strategy and products (services) that the company delivers to the marketplace. The GTI-based process of creation of strategic innovations is shown below (Fig. 2). The process of Strategic Innovation was created in cooperation with Dr. Paolo Mutti (Milan, Italy).

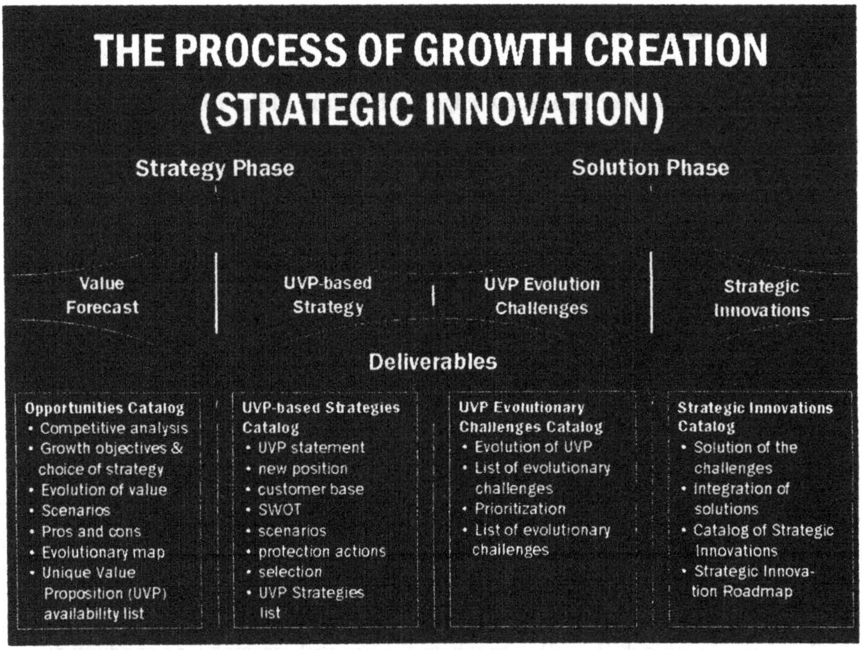


**Fig. 2.** The Process of Creating Strategic Innovations
(in cooperation with Dr. Paolo Mutti, Italy)

## 5 GTI: Looking into the Future

Although GTI has been developed for about 20 years and can answer many questions that were considered unanswerable even yesterday, it can and should be evolved further because there is no such a thing as a perfect theory; further improvement is always possible, and evolution is unstoppable. The author also recognizes that the progress can be achieved easier if pursued by a team of colleagues; so everyone interested in the subject of GTI is invited to contribute. Accomplishment of the following goals is considered critically important by the author.

1. Develop and continuously evolve the General Theory of Innovation (GTI) and its various applications.
2. Disseminate GTI throughout the society starting with primary education.
3. Seamlessly align GTI with other sciences and theories, including TRIZ Collaborate with societies and organizations with common interests.
4. Work diligently toward the situation where society will recognize Innovation as a distinct profession.
5. Standardize materials for and methods of preparing specialists pursuing the goal of becoming proficient in GTI and a variety of its applications.
6. Develop and introduce a process of certifying Innovators recognized by the global business and academic communities.

## 6 Acknowledgements

The author would like to acknowledge contribution of the following individuals to the creation and development of the General Theory of Innovation; without them the results that have been achieved would be absolutely impossible.

1. Genrich S. Altshuller, the late father of TRIZ, who encouraged the original research and provided a number of priceless suggestions;
2. Alexander A. Malinovsky (Bogdanov), the late father of Tectology (1920) whose work [18] prophetically preceded many of the results by Altshuller, Bertalanffy [19], and Wiener [20];
3. Dr. Lev L. Velikovich (Belarus), a close friend and mentor, who was the first to recognize the importance of GTI and encouraged the research;
4. Dr. John Terninko, a close friend and an intellectual sparring partner, who continuously challenged every aspect of GTI, including axioms and assumptions, thereby making GTI stronger;
5. Dr. Gaetano Cascini (Italy), a close friend and colleague; whose advice and opinion is always relevant and thoughtful;
6. Dr. Noel Leon (Mexico), who has invented the term of the General Theory of Innovation and generously presented it to the author;
7. Greg Frenklach (Israel) with whom the original research was conducted;
8. My countless TRIZ colleagues, whose works contributed to my upbringing and understanding of the world;

9. Last but not least; my family (wife Larisa Yezersky, our sons Geoffrey, Andrew and Alexander; my late father Abraham Yezersky, and my mother Fira Yezersky), whose unconditional love and unwavering support allowed me pursuing the goal of creating GTI.

## 7 References

1. Wikipedia, URL: http://en.wikipedia.org/wiki/Innovation.
2. Schumpeter, J. (1934), *The Theory of Economic Development* (Harvard University Press, Cambridge, Massachusetts).
3. W. Chan Kim, Renee Mauborgne, *Blue Ocean Strategy*, (HBS Press, Boston, 2005)
4. Alex F. Osborn, *Applied Imagination*, (Creative Education Foundation, 1993)
5. Fritz Zwicky, *Discovery, Invention, Research through the Morphological Approach*, (MacMillan Publishing Company, Toronto, 1969)
6. William J. J. Gordon, *Synectics*, (MacMillan Publishing Company, Toronto, 1971)
7. Edward De Bono, *Lateral Thinking*, (Harper Paperbacks, New York, 1973)
8. James M. Higgins, *101 Creative Problem Solving Techniques*, (New Management Publishing, 1994)
9. Michael MIchalko, *Thinkertoys*, (Ten Speed Press, Berkley, 1991)
10. Clayton M. Christensen, *The Innovator's Dilemma*, (HarperBusiness, New York, 2000)
11. Clayton M. Christensen, Michael E. Raynor, *The Innovator's Solution*, (HBS Press, Boston, 2003)
12. Clayton M. Christensen, Erik A. Roth, Scott D. Anthony, *Seeing What's Next*, (HBS Press, Boston, 2004)
13. Constantinos C. Markides, Paul A. Geroski, *Fast Second*, (Jossey-Bass, San Francisco, 2005
14. G. S. Altshuller, *Creativity As An Exact Science*, (Gordon and Breach, New York, 1988)
15. Genrich Altshuller, *The Innovation Algorithm*, (Technical Innovation Center, Worcester, 1999)
16. Steve Schoenherr, Recording Technology History, notes; Website URL: http://history.sandiego.edu/GEN/recording/notes.html
17. Glyn Davies, *A history of money*, (University of Wales Press, 1996) and also at Website URL: http://www.ex.ac.uk/~RDavies/arian/llyfr.html
18. Alexander A. Bogdanov (Malinovsky) *Tektologia*, (Economika Publishing House, Moscow, 1989, in Russian)
19. Ludwig von Bertalanffy, *General System Theory*, (George Braziler, New York, 1968)
20. Norbert Wiener, *Cybernetics*, (The Technology Press, New York, 1948)

# Locating Creativity in a Framework of Designing for Innovation

John S. Gero[1] and Udo Kannengiesser[2]
1 Krasnow Institute for Advanced Study and Volgenau School of Information Technology and Engineering, George Mason University, USA, and University of Technology, Sydney, Australia
john@johngero.com
2 NICTA, Australia
udo.kannengiesser@nicta.com.au

**Abstract.** This paper focuses on creativity in the process of designing as the foundation of potential innovations resulting from that process. Using an ontological framework that defines distinct stages in designing, it identifies the locations for creativity independently of their embodiment in human designers or computational tools. The paper shows that innovation, a consequence of creativity, can arise from a large variety of processes in designing.

## 1 Introduction

Innovation and creativity are terms that are often used synonymously. This is a result of them sharing the feature of establishing novelty, in the form of new products, systems, processes or organisations [1]. On the other hand, there is an important distinction between them: While creativity generates novel ideas, innovation puts these ideas into practice [2]. This paper focuses on creativity as the foundation and precursor of innovation. This is based on our belief that one of the highest potentials for computer support for design innovation comes from focussing on idea generation rather than only on downstream stages of idea realisation.

We define the notion of novelty, and consequently creativity, relative to the process of designing: Creativity occurs whenever a new design property is introduced for the first time in the ongoing design process, thus changing the state space of possible designs. While there are a number of environment-centric (socio-cultural, historic, technological, etc.) factors that determine whether the final product is ultimately accepted as an innovation, our definition of creativity represents a necessary precondition. We will use an ontological framework to locate creativity at various places in the process of designing.

*Please use the following format when citing this chapter:*

Gero, J. S., Kannengiesser, U., 2007, in IFIP International Federation for Information Processing, Volume 250, Trends in Computer Aided Innovation, ed. León-Rovira, N., (Boston: Springer), pp. 57-66.

## 2 Activities in Creative Designing

Research in creative designing can be separated into two strands. One strand of research is concerned with developing computational processes that can extend the design state space. Here, five classes of processes have been suggested [3]: emergence, analogy, combination, mutation and first principles. Another strand of research deals with human behaviour involved in changing the state space of possible designs. Here, creativity is viewed as part of more general design behaviour including cognitive and physical activities carried out by the designer.

A paradigm that characterises designing in terms of the activities performed by human designers is situatedness [4, 5, 6]. From the situated perspective, designers perform actions in order to change their environment. By observing and interpreting the results of their actions, they then decide on new actions to be executed. This means that the designers' concepts may change according to what they are "seeing", which itself is a function of what they have done. One may speak of an "interaction of making and seeing" [7] that strongly determines the course of designing by modifying the design state space.

A number of activities related to potential design creativity have been studied that fit with the notion of situatedness. Most of them can be grouped into two classes: interpretation and reflection. Interpretation of design sketches can lead to unexpected discoveries and the invention of new issues or requirements during the design process [7, 8]. Reflection can bring about similar effects, but is based on internal memory construction rather than external sketches [9]. Both interpretation and reflection are not simple mappings driven by either pure "data push" (replicating previous sketches or memories) or pure "expectation pull" (replicating current expectations). They are rather interactive combinations of both "push" and "pull", and new design concepts emerge from this interaction [10]. Interpretation and reflection are the source of modifications in the design state space. They are orthogonally related to Gero's [3] five computational processes; i.e., any of these processes may occur in any of the two classes of design activities.

## 3 An Ontological Framework of Designing

### 3.1 The FBS Ontology

Gero [11] has proposed an ontology of design objects that provides three high-level categories for the properties of an object: function, behaviour and structure.

*Function* (F) of an object is defined as its teleology, i.e. "what the object is for". An example is the function "to wake someone up" that humans generally ascribe to the behaviour of an alarm clock.

*Behaviour* (B) of an object is defined as the attributes that are derived or expected to be derived from its structure, i.e. "what the object does". An example of behaviour is "weight", which can be derived directly from a physical object's structure properties of material and spatial dimensions.

*Structure* (S) of an object is defined as its components and their relationships, i.e. "what the object consists of". It represents the object's "building blocks" that can be directly created or modified by the designer. Examples include molecule structures, mechanical structures, floor and wall structures, and spatial structures.

## 3.2 The Situated FBS Framework

Gero and Kannengiesser [12] have proposed a three-world model of design interactions, Figure 1(a). The *external world* is composed of representations outside the design agent. The *interpreted world* is built up inside the design agent in terms of sensory experiences, percepts and concepts. It is the internal representation of that part of the external world that the design agent interacts with. The *expected world* is the world imagined actions of the design agent will produce. It is the environment in which the effects of actions are predicted according to current goals and interpretations of the current state of the world.

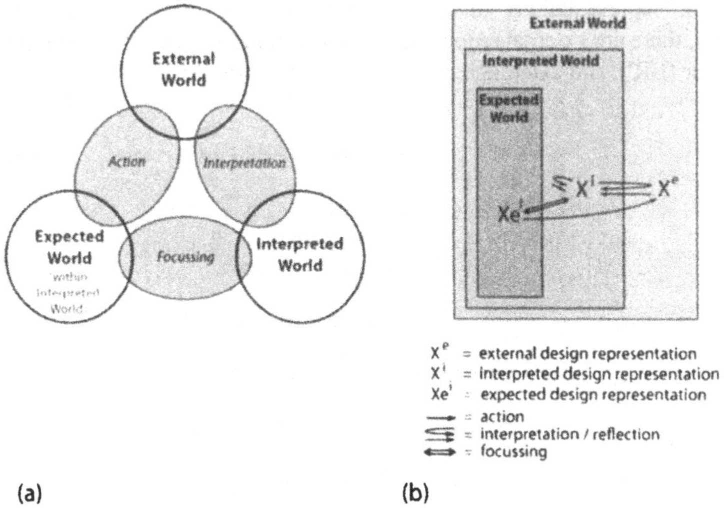


**Fig. 1.** Three interacting worlds: (a) general model, (b) specialised model for design representations (after [12]).

These three worlds are linked together by three classes of connections. *Interpretation* transforms variables which are sensed in the external world into the interpretations of sensory experiences, percepts and concepts that compose the interpreted world. *Focussing* takes some aspects of the interpreted world, and uses them as goals for the expected world that then become the basis for the suggestion of actions. These actions are expected to produce states in the external world that reach the goals. *Action* is an effect which brings about a change in the external world according to the goals in the expected world.

Figure 1(b) specialises this model by nesting the three worlds and adding general classes of design representations as well as the activity of reflection [9]. The set of expected design representations ($Xe^i$) corresponds to the notion of a design state space, i.e. the state space of all possible designs that satisfy the set of requirements. This state space can be modified during the process of designing by transferring new interpreted design representations ($X^i$) into the expected world and/or transferring some of the expected design representations ($Xe^i$) out of the expected world. This leads to changes in external design representations ($X^e$), which may then be used as a basis for re-interpretation changing the interpreted world. Novel interpreted design representations ($X^i$) may also be the result of reflection, which can be viewed as a process of interaction among design representations within the interpreted world rather than across the interpreted and the external world. Both interpretation and reflection are represented as push-pull activities.

# 4 Locating Creativity in the Situated FBS Framework

Gero and Kannengiesser's [12] situated FBS framework, Figure 2, combines the FBS ontology with the three-world model. Here, the variable X in Figure 1(b) is replaced with the more specific representations F, B and S. The situated FBS framework also uses explicit representations of external requirements given to the designer. Specifically, there are external requirements on function ($FR^e$), external requirements on behaviour ($BR^e$), and external requirements on structure ($SR^e$).

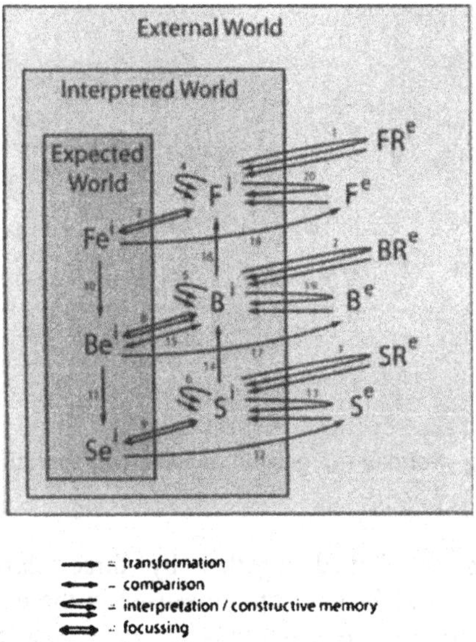


**Fig. 2.** The situated FBS framework (after [12]).

The situated FBS framework specifies a set of 20 labelled activities that can be mapped them onto eight fundamental design steps [11]. The remainder of this Section will identify potential locations of creativity in each of these steps.

### 4.1 Formulation and Creativity

Formulation comprises activities 1 to 10, producing an initial state space of potential design structure, behaviour and function, based on externally given requirements or internal reflection. This design step can already be viewed as a creative act, as it produces a set of design concepts for the first time in the design process. It is based on interpretation and reflection as the generators of new design concepts.

Interpretation (activities 1, 2 and 3) is not a simple reproduction of what is given to the designer. Any two designers will be likely to interpret the same requirements differently, depending on their individual experiences. As a result, they produce different concepts of function, behaviour and structure. This is often based on ambiguities in the requirements given to the designer, especially for function that is often expressed in informal ways.

Reflection (activities 4, 5 and 6) produces additional, implicit requirements that have not been explicitly given to the designer. Cross [13] provides the example of an expert engineer performing the task of designing a device that allows fastening and carrying a backpack on a mountain bike. Based on the engineer's personal experience as a cyclist, an implicit constraint on the structure of the device was constructed, namely to select its location (and thus the location of the backpack) as low as possible for better riding control.

The transformation of function into behaviour (activity 10) can be supported by analogy processes. For example, behaviours of a flexible manufacturing system (FMS) have been specified based on similarities with the functions of an air traffic control (ATC) system [14], Figure 3. Here, functions such as "prevent collisions" and "ensure movements according to the pre-defined plan" are general goals of both systems. Behaviours of one system can then be mapped onto the other.

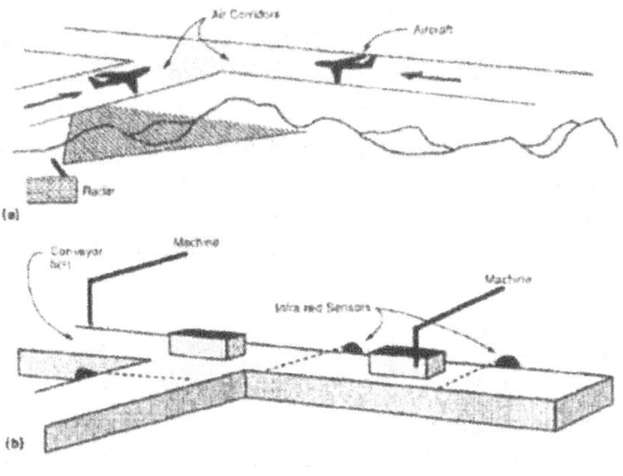


**Fig. 3.** Key features of the domains of (a) air traffic control and (b) flexible manufacturing (taken from [14]).

### 4.2 Synthesis and Creativity

Synthesis generates a design solution in terms of a specific instance of structure (activity 11), and subsequently produces an external representation of that instance (activity 12). Synthesis does not modify the design state space, as it uses the concepts produced by formulation without adding new ones, operating only on the values of structure variables rather than on the structure variables themselves. This view excludes a direct association of creativity with the synthesis step. The connection between synthesis and creativity lies in the interpretation of synthesised structure.

### 4.3 Analysis and Creativity

Analysis derives behaviour from synthesised design structure. In the situated FBS framework, this design step is composed of activities 13 and 14. Creativity is generally not desired in analysis, as the primary purpose of this design step is to prepare for the evaluation of the design solution against the pre-determined set of criteria, i.e. within the bounds of the current design state space. However, it is often during the act of analysis that new design issues and ideas emerge that may lead to additions to the design state space. One driver is the interpretation of structure (activity 13), providing the potential for the process of emergence to occur. Examples presented in [15] are concerned with the emergence of shapes through re-representation of external structure. The wide range of drafting tools available may support re-representations, such as 2D-, 3D-, "fly-through" and simulation models, each of which can enable the discovery of specific types of design issues.

Emergence during the analysis step is not limited to the structure level. New behaviour often emerges when using computational analysis tools that derive more performances from a given structure than necessary for the currently intended evaluation (activity 14). This is more correctly called supervenience although it is generally taken to be a form of emergence. Take the designer of a material flow system, interested in evaluating the speed of that system. The simulation tool used for the analysis calculates not only speed, but also machine utilisation. This can be seen as the generation of an additional behaviour to be considered in the design state space. Here, creativity is located in the interaction between the designer and the tool.

### 4.4 Evaluation and Creativity

Evaluation assesses the design solution on the basis of the formulated criteria, i.e. by comparison of the behaviour derived from the design solution with the expected behaviour (activity 15). No creativity is involved here.

### 4.5 Documentation and Creativity

Documentation produces an external representation of the final design solution for purposes of communicating that solution (activities 12, 17 and 18) to a builder, manufacturer or implementer. Creativity plays no role here, because the process of design generation is terminated prior to documentation.

## 4.6 Reformulation Type 1 and Creativity

Reformulation type 1 modifies the structure state space (activity 9). It is based on creative processes that may produce new structure variables. Any of three classes of input is needed for these processes: external requirements on structure, external representations of existing design structure, and interpreted structure representations.

The interpretation of external requirements on structure (activity 3) can bring about new structure variables in two cases. In the first case, modified external requirements are given to the designer after commencement of the design process. Here, the creativity is located in the environment in which the design process is carried out rather than in the design process itself. In the second case, the same external requirements as given at the outset of the design process are interpreted differently. This locates the creativity in the process of interpretation.

The interpretation of external design structure (activity 13) can involve creative processes that generate new structure variables. One example is the process of analogy. Here, an external structure is a source design exhibiting identical behaviour as the current (target) design. The matching and then mapping of the source structure onto the target structure is the creative element of interpretation.

Reflection on interpreted structure (activity 6) constructs new structure variables without the use of external representations. This creative activity is frequently used across the design disciplines to lead the design process into new directions [9]. Reflection on structure is the best-studied type of the reflection processes. For example, the domain of drug design uses reflection on structure to incrementally modify molecules to generate new drugs that meet specified behaviours. Here, reflection is often implemented as crossover (or combination) and mutation processes, embedded in design methodologies using Genetic Algorithms (GAs). GAs take inspiration from biological evolution by applying crossover and mutation operators over populations of candidate structures. Candidates are selected using a fitness function that measures individual performances that correspond to the values of a given behaviour.

## 4.7 Reformulation Type 2 and Creativity

Reformulation type 2 modifies the behaviour state space (activity 8). It is based on creative processes that produce new behaviour variables. Any of four classes of input is needed for these processes: external requirements on behaviour, external representations of design behaviour, interpreted behaviour representations, and interpreted structure representations.

The interpretation of external requirements on behaviour (activity 2) can produce new behaviour variables in an analogous way as in reformulation type 1. Creativity is located either outside the system via modified requirements or inside the system via modified interpretation.

The interpretation of external design behaviour (activity 19) can produce new behaviours to alter the behaviour state space. This activity often uses emergence to reconceptualise current behaviours, which has mainly been studied in software requirements engineering [16].

Reflection on interpreted behaviour (activity 5) constructs new behaviour variables without the use of any external representations. It reconceptualises current behaviours in a similar way as interpretation. Its underlying mechanism can be assumed to be emergence, although little research has been done here.

The derivation of additional behaviour from structure (activity 14) can drive reformulation type 2 in the same way as described for the analysis step. Creativity here is fostered by the interaction between designers and computational tools. Another way to derive new behaviour from structure is via analogy. Consider the following example: A lamp designer, after coming home from work, looks at a book that still lies half open on his bedside table, Figure 4(a). He becomes aware that the front and back covers of the book can be thought of as defining an arched aperture that can be expanded or reduced by opening or closing the covers. This can be viewed as behaviour derived from an abstraction of the book's structure. The designer realises that in the lighting domain this behaviour could be used as a physical mechanism to fulfil the function of dimming light. He finds this idea interesting and decides to change the design of a desk lamp he is currently working on by implementing this physical dimming mechanism instead of the originally intended electrical one. Figure 4(b) shows a CAD model of the new lamp. It incorporates behaviour and structure features of the book as the source design.

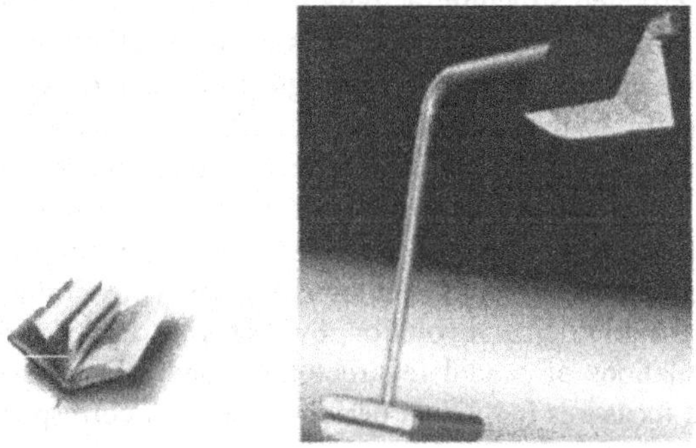

(a) (b)

**Fig. 4.** Example of analogy in interpretation: (a) a half-open book as a source structure, (b) the desk lamp (model "Hamlet") as the target structure (Source: Stimulo Design).

## 4.8 Reformulation Type 3 and Creativity

Reformulation type 3 modifies the function state space (activity 7). It is based on creative processes that produce new function variables. Any of four classes of input is needed for these processes: external requirements on function, external representations of design function, interpreted function representations, and interpreted behaviour representations.

The interpretation of external requirements on function (activity 1) can produce new function variables in an analogous way as in reformulation type 1 and 2. Creativity is located either outside the system via modified requirements or inside the system via modified interpretation.

The interpretation of external design function (activity 20) can produce new functions to alter the function state space. Maiden et al. [17] have described how designers of air traffic management software combined existing software functions into new ones during a creativity workshop. Here, the two functions "to allow air traffic controllers to maintain an accurate mental model of the air space" and "to offer new types of situational display to air traffic controllers" were combined into the new function "to allow air traffic controllers to rewind and fast-forward aircraft movements to develop their mental models of the air space before taking responsibilities for decisions that they will make".

Reflection on interpreted function (activity 4) constructs new functions without the use of any external representations. Little research has been done in this area.

The derivation of function from behaviour (activity 16) can result in new functions to be incorporated in the design. Schön [18] has presented the example of Scotch Tape, whose initial function was "to mend books". However, the people buying this product invented additional functions, such as "to wrap packages" and "to curl hair". These functions have been likely to be derived from the behaviour of removable adhesiveness. They subsequently led to adaptations of the product to different applications such as packaging and hair curling.

## 5 Conclusion

We have shown in this paper that the situated FBS framework provides an ontological basis for locating creativity at various stages in designing. While the three classes of reformulation are the design steps that include most creative activities, the framework can locate creativity also at other places in the design process.

Most importantly, our framework provides the foundations for developing better computational tools to support creativity and innovation. This is based on the independence of an ontological view of creativity with respect to its embodiment. All creative processes and activities in designing can be located in human designers, their tools and the interaction between the designers and the tools.

Innovation is rooted in creativity. Whilst creativity is the process of generating new designs, innovation is the process of putting novel designs into practice. It requires a knowledge of what is novel in terms of function, behavior and structure to be able to demonstrate that the design is novel in at least one of those dimensions. An ontological framework of designing allows both the locus of creativity to be determined and the support for innovation to be articulated.

## 6 Acknowledgements

This research is supported by a grant from the Australian Research Council, grant no. DP0559885 – Situated Design Computing. It was carried out at the Key Centre of Design Computing and Cognition, University of Sydney. The authors wish to thank the company Stimulo Design for providing the example of the "Hamlet" desk lamp.

## 7 References

1. J. Fagerberg, D.C. Mowery and R.R. Nelson, *The Oxford Handbook of Innovation* (Oxford University Press, Oxford, 2005).
2. D. Gurteen, Knowledge, creativity and innovation, *Journal of Knowledge Management* **2**(1), 5-13 (1998).
3. J.S. Gero, Creativity, emergence and evolution in design, *Knowledge-Based Systems* **9**(7), 435-448 (1996).
4. J. Dewey, The reflex arc concept in psychology, *Psychological Review* **3**, 357-370 (1896 reprinted in 1981).
5. F.C. Bartlett, *Remembering: A Study in Experimental and Social Psychology* (Cambridge University Press, Cambridge, 1932 reprinted in 1977).
6. W.J. Clancey, *Situated Cognition: On Human Knowledge and Computer Representations* (Cambridge University Press, Cambridge, 1997).
7. D.A. Schön and G. Wiggins, Kinds of seeing and their functions in designing, *Design Studies* **13**(2), 135-156 (1992).
8. M. Suwa, J.S. Gero and T. Purcell, Unexpected discoveries and s-inventions of design requirements: A key to creative designs, in: *Computational Models of Creative Design IV*, edited by J.S. Gero and M.L. Maher (Key Centre of Design Computing and Cognition, University of Sydney, Sydney, Australia, 1999), pp. 297-320.
9. D.A. Schön, *Educating the Reflective Practitioner: Toward a New Design for Teaching and Learning in the Professions* (Jossey-Bass Publishers, San Francisco, 1987).
10. J.S. Gero and H. Fujii, A computational framework for concept formation for a situated design agent, *Knowledge-Based Systems* **13**(6), 361-368 (2000).
11. J.S. Gero, Design prototypes: A knowledge representation schema for design, *AI Magazine* **11**(4), 26-36 (1990).
12. J.S. Gero and U. Kannengiesser, The situated function-behaviour-structure framework, *Design Studies* **25**(4), 373-391 (2004).
13. N. Cross, Creative cognition in design: Processes of exceptional designers, in: *Creativity and Cognition'02*, edited by T. Hewett and T. Kavanagh (ACM Press, New York, 2002), pp. 14-19.
14. N.A. Maiden and A.G. Sutcliffe, Exploiting reusable specifications through analogy, *Communications of the ACM* **35**(4), 55-63 (1992).
15. H.J. Jun and J.S. Gero, Representation, re-representation and emergence in collaborative computer-aided design, in: *Preprints Formal Aspects of Collaborative Computer-Aided Design*, edited by M.L. Maher, J.S. Gero and F. Sudweeks (Key Centre of Design Computing and Cognition, University of Sydney, Australia, 1997), pp. 303-320.
16. L. Nguyen and P.A. Swatman, Managing the requirements engineering process, *Requirements Engineering* **8**(1), 55-68 (2003).
17. N.A. Maiden, A. Gizikis and S. Robertson, Provoking creativity: Imagine what your requirements could be like, *IEEE Software* **21**(5), 68-75 (2004).
18. D.A. Schön, 1983, *The Reflective Practitioner: How Professionals Think in Action* (Harper Collins, New York, 1983).

# Research and Implementation of Product Functional Design Based on Effect

Cao Guozhong, Tan Runhua and Lian Benning
School of Mechanical Engineering, Hebei University of Technology,
Tianjin, P.R. China
cgzghx@163.com

**Abstract.** Functional design is a process includes function understanding, function recognition, function representation, function modeling and function-to-structure mapping. The effect in TRIZ is extended by multi-pole effect model and effect chain modes, and then the functions and flows in TRIZ are reclassified. The combinative relationships among functions are discussed, and the function modeling based on effect is proposed. The principle structure is introduced, and the combined rules are presented. The functional design process model for function modeling and function-to-structure mapping is proposed, and the computer-aided functional design software is developed. A design example is presented to demonstrate the proposed functional design methodology.

**Keywords.** Functional Design, Effect, Principle Structure

## 1 Introduction

Functional design, which plays the central role in ensuring design quality and product innovation, is a well-researched and active field of engineering study.

There are various, often conflicting, definitions of function in the literature; no universally accepted definition is currently known, such as, designer's purpose [1,2], intended behavior [3], an effect on the environment of the product [4], a description of behavior recognized by a human through abstraction in order to utilize it [5] or a relationship between inputs and outputs, aiming to achieve the designer's purpose [6]. Clearly, each of these definitions has some aspects of worth, yet none are comprehensive enough to capture the fullness of definition that is desired.

Researchers have recognized the importance of a common vocabulary for broader issues of design. Pahl and Beitz [6] list five generally valid functions and three types of flows at a very high level of abstraction. Collins et al.[7] develop a list of 105 mechanical functions, which are limited to helicopter systems and do not utilize any classification scheme. TRIZ describes all mechanical design with a set of 30

*Please use the following format when citing this chapter:*

Guozhong, C., Runhua, T., Benning, L., 2007, in IFIP International Federation for Information Processing, Volume 250, Trends in Computer Aided Innovation, ed. León-Rovira, N., (Boston: Springer), pp. 67-76.

functional descriptions [8]. Malmqvist et al. [9] compare TRIZ with the Pahl and Beitz methodology and note that the detailed vocabulary of TRIZ would benefit from a more carefully structured class hierarchy using the Pahl and Beitz functions at the highest level.

Functional composition has been given considerable attention by researchers. Systematic design [6] characterizes an overall function can be adequately defined in terms of quantified inputs and outputs, and use input/output flow analysis and synthesis method to model a function structure. Axiomatic design [10] states that functional decomposition should be viewed as a zigzag mapping between functional requirements in a functional domain and design parameters in a physical domain. Struges et al. [11] present a scheme that is able to specify both vertical and horizontal functional relationships by functional block diagram. Chen [12] defines a general functional representation structure in terms of category, level and layer. Top-down decomposition methods provide a means for dividing a general function into sub-functions, but they can not describe the key characteristics of systems.

One of the most well-known functional design frameworks is that of Pahl & Beitz [6], i.e., systematic approach, which model the overall function and decompose it into sub-functions operating on the flows of energy, material, and signals. Umeda et al. [13] proposed a Function-Behavior-State (FBS) modeler that reasons about function by means of two approaches: causal decomposition and task decomposition. Deng et al. [14] devised a dual-step function-environment-behaviour-structure (FEBS) model There are other similar approaches for functional models, for example, Qian and Gero's [1] FBS Path, and Prabhakar and Goel's [2] ESBF model.

During functional design the design knowledge and technologies in multiple different domains may be employed, and complicated developing activities. Although there are now some general methodologies dealing with functions or reasoning about functions, at present time, satisfying solution to both issues has not yet arrived [15].

Aiming at solving some crucial issues discussed above, this paper constructs a functional design process model for function modeling and function-to-structure mapping, which is based on multi-pole effect model, effect chain modes and combined rules of principle structure.

## 2 Effect

Effect is one of the knowledge base tools in TRIZ. By the analysis of hundreds of thousands patents, effects are emerged from the relevance between functions delivered by a design product described in a patent and a principle used in the product [16]. An Effects is an input and output relationship that combines the laws of science including physics, mathematics, chemistry and geometry, and their corresponding engineering applications, which helps to bridge the gap between science and engineering and is good for generating solutions of high levels. However, few literatures on effect exist in TRIZ monographs and few researches on effect exist in international, for example, only the concept, sort and use of effect are introduced in literature [17-19]. There is no literature on functional design based on effect.

### 2.1 Effect Model

Effects can be characterized by its input, output relations [20]. An effect has an input and output flow, which is called as basic effect, thus the effect model has two poles [33], as shown in Fig. 1(a). The two-pole effect model is extended in this paper. Most transitions from input to output with effect are controlled by auxiliary flow, so the controllable effect should be denoted with three poles, as shown in Fig. 1(b). The control flow specifies the factors that can be manipulated to change the output intensity of an effect. Generally speaking, an effect may have multiple input flows, output flows or control flows, so the effect has multiple input poles, output poles or control poles, as shown in Fig. 1(c).

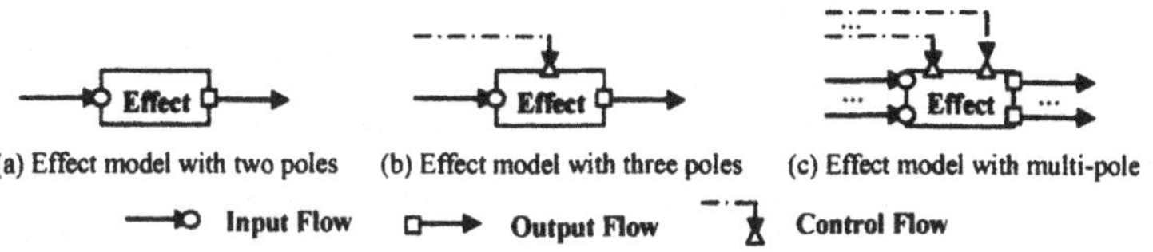


**Fig. 1.** Effect model

### 2.2 Effect Chain and Effect Mode

By the analysis of patents, the mapping relationship between function and effect is confirmed. The function fulfilled directly by effect is called as function unit. But, it is difficult for complex function to perform it by an effect. The effects need to be linked sequentially into chains through their input, output or control ports and compatibility of adjacent effects. Effect modes are the basic coupling manners of effect chains, as shown in the following:

- Serial effect mode: composed of several effects occurring in sequence, as shown in Fig. 2(a).
- Parallel effect mode: composed of several effects occurring at same time, as shown in Fig. 2(b).
- Ring effect mode: composed of several effects a set of effects which the input of a former effect is the output of a latter effect, as shown in Fig. 2(c).
- Control effect mode: the internal characteristic of an effect can be controlled by other effects in order to control output of the effect, as shown in Fig. 2(d).
- Combined effect mode: is composed of several above effects modes.

The effects can be linked into effect chain by using effect modes. During the transformation from input flow to output flow by effect modes, there are three methods, namely, method of exhaustion, method of minimal path length and method of consistent degree. The effect chain can help engineers achieve breakthrough innovation by proposing new and unexpected variations in producing a specific output.

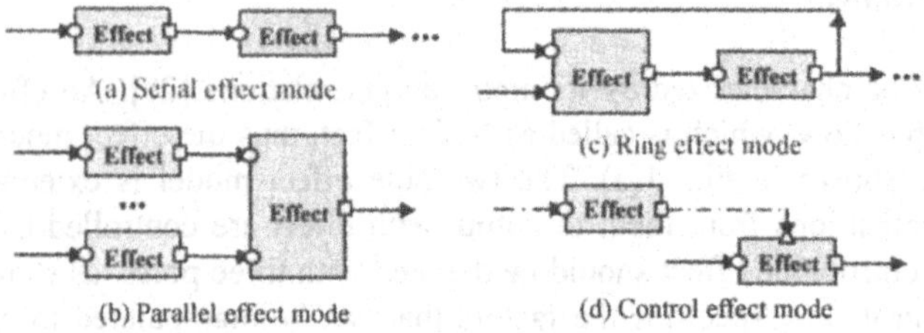


**Fig. 2.** Effect modes

# 3 Function

## 3.1 Function Set

Function is a statement to describe the transformation between inputs and outputs, aiming to achieve the designer's purpose. A function consists of commonly input/output part, and operation part. The input and output entities are referred to as functional flows. The operations are referred to as behaviors. Function is expressed as noun-verb-noun, which is different from the accustomed expression of verb-noun. Thus the function and effect have a consistent pattern of expression, and the horizontal interaction relationships among sub-functions can be confirmed and verified by compatibility of adjacent inputs and outputs.

Subsuming the classification schemes discussed above, the 30 functions in TRIZ are expanded and reclassified and the standard set of functions is presented, which includes a behavior (verb) set and a flow (noun) set, as shown in Table 1 and 2.

**Table 1.** Short list of behavior set

| Behavior | Sub-behavior |
|---|---|
| Create | Synthesize, Produce |
| Change | Increase, Decrease, Convert, Deform, Control |
| Combine | Mix, Embed, Assemble, connect |
| Separate | Disassemble, Decompose, Dry, Clean |
| Accumulate | Absorb, Store, Concentrate |
| Move | Move, Transfer, Rotate, Vibrate, Lift, Orient |
| Measure | Determine, Detect, Measure |
| Preserve | Preserve, Prevent, Stabilize |
| Eliminate | Destroy, Remove |

**Table 2.** Short list of flow set

| Flow | Sub-flow |
|---|---|
| Material | Solid, Liquid, Gas, Geometric objects, Loose Substances, Porous Substances, Particles, Plasma, Chemical Compounds |
| Energy | Forces, Motion, Deformation, Thermal Energy, Mechanical and Sound Waves, Electric Field, Magnetic Field, Nuclear Energy, Electromagnetic Waves or Light |
| Parameters | Surfaces Parameters, Geometric Parameters, Fluids Parameters, Concentration Parameters, Forces Parameters, Motion and Vibration Parameters, Thermal Parameters, Electric field Parameters, Magnetic field Parameters, etc. |

## 3.2 Function Model

Products are defined by the overall functions. The overall function can be broken down into several sub-functions. The aggregation of functions and their relations is called function model. There are two types of relations between functions.

One is the relation generated through function decomposition, called hierarchical relation, which illustrates the hierarchical relationships among sub-functions, and discovers the hierarchical design purposes. There are two logical relationships: conjunction and disjunction.

- Conjunction structure: the parent function is achieved by several corporate sub-functions, as shown in Fig. 3(a).
- Disjunction structure: the parent function is achieved by only one of several sub-functions according to condition, as shown in Fig. 3(b).

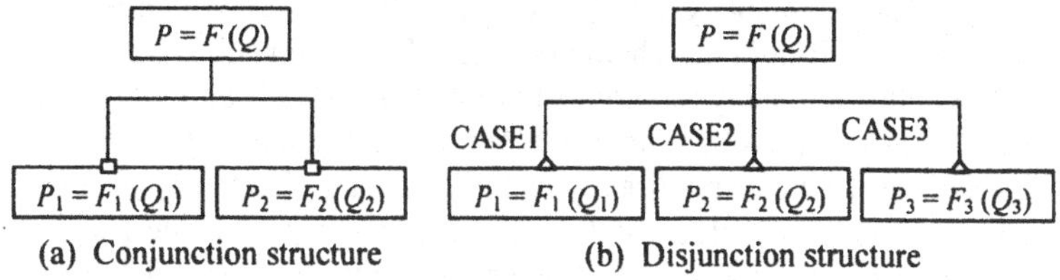


**Fig. 3.** The hierarchical relationships among sub-functions

Another is lateral relation if the input flow of one function is the output flow of the other, which indicates the interaction relationships among active sub-functions at the related level of functional model, reflecting logically and physically possible or useful associations of the sub-functions. There are four logical relationships: serial, parallel, ring and control.

- Serial structure: the parent function is achieved by sub-functions occurring in sequence, as shown in Fig. 4(a).
- Parallel structure: the parent function is achieved by sub-functions happening at the same time, as shown in Fig. 4(b).
- Ring structure: the parent function is achieved by sub-functions in which the input of a former function is the output of a latter function, as shown in Fig. 4(c).
- Control structure: the parent function is achieved by several sub-functions in which a function is controlled by other function, as shown in Fig. 4(d).

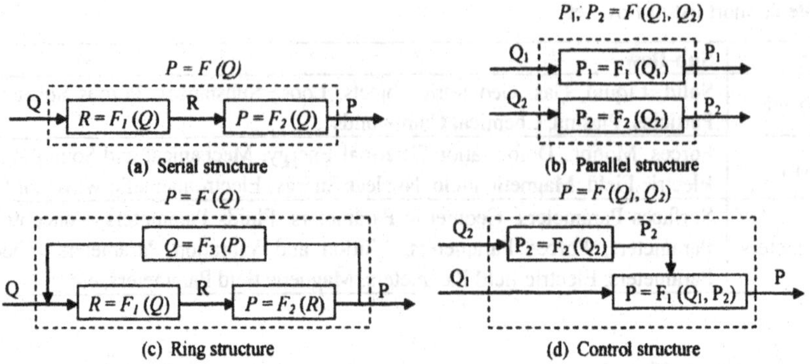


**Fig. 4.** The lateral relationships among sub-functions

## 4 Structure

In conceptual design structure is not concrete structure, but principle structure, which specifies what elements the structure is composed of, what the attributes of the elements are and how they are related.

Any structure is composed of elements. An element in the structure that cannot be divided is called a primitive element. A primitive element can be either a physical or logical entity. Some elements group together and form a sub-structure or structural element that has well-defined characteristics. The interactions of elements are physical interconnection using topological or geometrical data and logical processes.

Structure is a physical embodiment of effect and the change of structure from one state to another must be caused directly or indirectly by effects. By means of patent analysis, the structures, which show how the effect is used for the performance of function transitions, can be acquired. A structure contains structural feature, such as what elements the design is composed of, what the attributes of the elements are and how they are related. An effect can have varied realizing structure, and a structure can be used for varied effects.

## 5 Functional Design Based on Effect

The process of functional design can be seen as transforming a functional representation to a design description or physical representation through function, effect and Structure, as shown in Fig. 5.

- Design begins with the analysis of functional requirements (Q), and then determines the overall function of product (f) in function set.
- According to the inputs and outputs of function, the effect chain can be set up based on effect modes.
- The four logical relationships: serial, parallel, ring and control in effect chain are consistent with them in function model. By the mapping relationship between effect and function, the effects in chain are replaced by relevant function .Thus the

relationship among sub-functions can be confirmed and the function model is constructed.

- The function model algebra system and the Boolean algebra system are isomorphic, so the function model can be simplified by Boolean calculation of sub-functions.
- By the mapping relationship between effect and structure, each sub-function in function model has its structure, and the product structure is established.

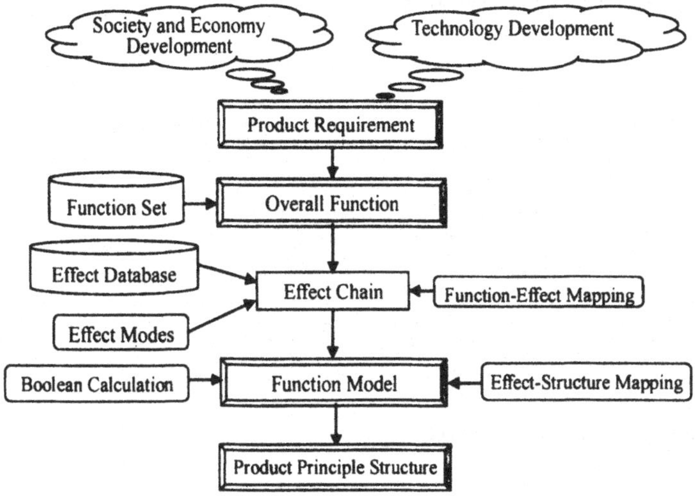


**Fig. 5.** Functional design process model based on effect

Based on the functional design process, the computer-aided functional design Software (Effect for Functional Design, E2F Design) (in Chinese) is developed by cobweb model and v-model process of software development in Fusion, and is further built by using VC++ and XML. This platform can support function-oriented design in a smooth way.

## 6 Case Study

Pill is a kind of good form of Chinese traditional medicine, but it can not be produced by Western medicine facility for its process and physics characteristic. The present condition is long process, high energy consume and great labor intension, so it is important to develop continue forming and shorten process to meet the need of modern times.

The overall function of granulator system can be modeled in Fig. 6, whose inputs are powder (medicinal powder) and liquid (cementing liquid), and whose outputs are sphericity and particle (pill). According to the known inputs and outputs, search for the effects in effect database of E2F Design software. The effects can be automatically linked into effect chains by using effect modes. Fig. 7 shows the part of effect chains of granulator system. Fig. 8 shows the function model of granulator system based on effect chain in Fig. 7(c). Fig. 9 shows the principle structure of granulator system.

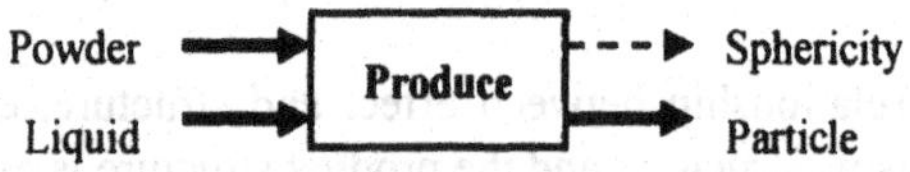


**Fig. 6.** Overall function of granulator system

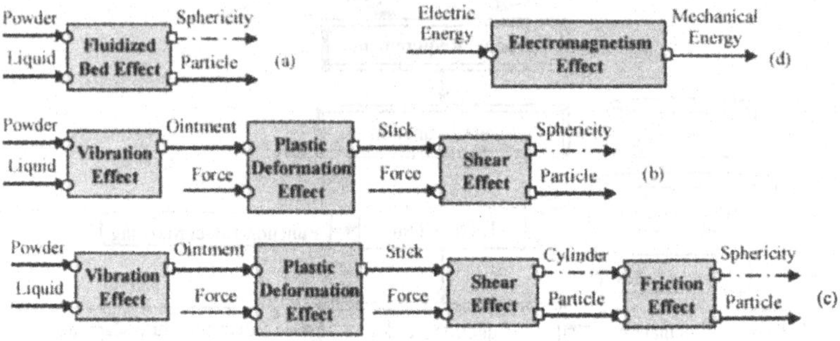


**Fig. 7.** Effect chains of granulator system

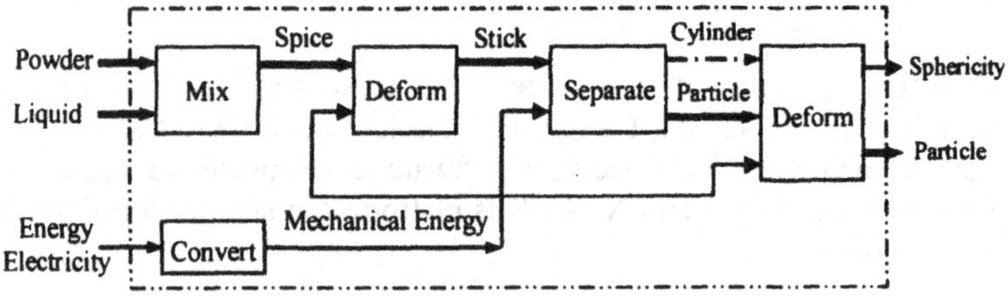


**Fig. 8.** Function model of granulator system

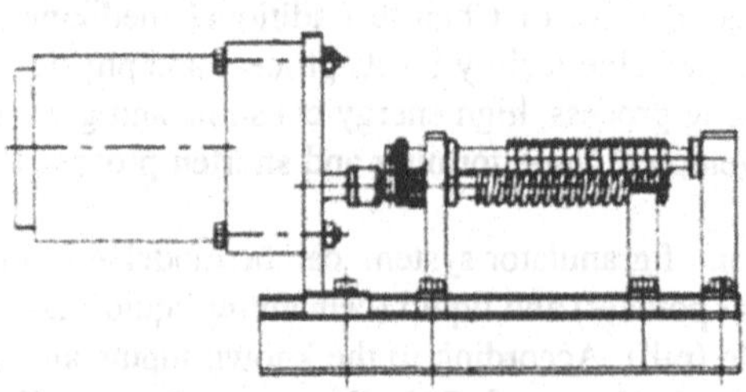

**Fig. 9.** Principle structure of granulator system

## 7 Conclusion

Functional design, which includes function understanding, function recognition, function representation, function modeling and function-to-structure mapping, plays the central role in ensuring design quality and product innovation. How to reason effectively and describe completely function model and how to map from function to structure are crucial issues in conceptual design phase.

This paper proposes a product functional design approach based on effect by the followings:

- The multi-pole effect model and effect modes for effect chain are created to extend effect in TRIZ;
- The functions and flows in TRIZ are reclassified and the standard function set is presented;
- The combinative relationships among functions are discussed and the function modeling based on effect is proposed;
- The functional design process model for function modeling and function-to-structure mapping is proposed, and the computer-aided functional design software is developed.
- A design example for conceptual design of air exhaust valve in car is presented to demonstrate the proposed functional design process and prove that the method is feasible.

## 8 Acknowledgements

The research is supported in part by the Chinese Natural Science Foundation under Grant Numbers 50675059 and Chinese national 863 planning project under Grant Number 2006AA04Z109. No part of this paper represents the views and opinions of any of the sponsors mentioned above.

## 9 References

1. L. Qian, and J.S. Gero, Function-Behavior-Structure Paths and their Role in Analogy-based Design, AIEDAM, 10, 289-312 (1996).
2. G. Prabhakar, and A. Goel, A functional modeling for adaptive design of devices in new environments, Artificial Intelligence in Engineering Journal (Special Issue), 12(4), 417-444 (1998).
3. Y. Shimomura, H. Takeda and et al., Representation of Design Object based on the Functional Evolution Process Model. DTM'95-ASME (1995).
4. B. Chandrasekaran and H. Kaindl, Representing Functional Requirements and User-system Interactions. AAAI Workshop on Modeling and Reasoning about Function, pp.78-84 (1996).
5. Y. Umeda and T. Tomiyama, FBS Modeling: Modeling Scheme of Function for Conceptual Design, Workshop on Qualitative Reasoning about Phys. Systems, Amsterdam, pp.271-278 (1995).
6. G. Pahl and W. Beitz, Engineering Design – A Systematic approach, The 2nd Edition, Springer-Verlag London (1996).

7. J. Collins, B. Hagan and H. Bratt, The Failure-Experience Matrix - a Useful Design Tool, Transactions of the ASME, Series B, Journal of Engineering in Industry, 98, 1074-1079 (1976).
8. G. Altshuller, Creativity as an Exact Science, Gordon and Branch Publishers, Luxembourg (1984).
9. J. Malmqvist, R. Axelsson, and M. Johansson, Comparative Analysis of the Theory of Inventive Problem Solving and the Systematic Approach of Pahl and Beitz, Proceedings of the 1996 ASME Design Engineering Technical Conference and Computers in Engineering Conference, Irvine, CA (1996).
10. N.P. Suh, The Principle of Design. New York: Oxford University Press (1990).
11. R.H. Sturges, Computational model for conceptual design based on extended function logic, Artificial Intelligent for Engineering Design Analysis and Manufacturing, 10, 255-274 (1996).
12. B.S. Chen, Characterization and Representation of Functional Requirements and Functional Tolerancing for Concurrent Design and Manufacturing [Ph. D. Thesis], Ohio State University, USA, (1994).
13. Y. Umeda, M. Ishii, M. Yoshioka, et al., Supporting conceptual design based on the function-behavior-state modeler, Artificial Intelligence for Engineering Design, Analysis and Manufacturing: Aiedam, 10 (4), 275-288 (1996).
14. Y.M. Deng, S.B. Tor, and G.A. Britton, Abstracting and exploring functional design information for conceptual product design, Engineering with Computers, 16, 36-52 (2000).
15. D. Tate, A Roadmap for Decomposition: Activities, Theories, and Tools for System Design [Ph.D Thesis], Department of mechanical Engineering, MIT, Cambridge, MA USA, (1999).
16. G. Altshuller, The Innovation Algorithm, TRIZ, Systematic Innovation and Technical Creativity, Technical Innovation Center, INC, Worcester (1999).
17. F. Gregory, Classifying the Technical Effects, TRIZ Journal, 3, http://www.triz - journal.com (1998).
18. Kalevi R. Eleven Uses of Effects. TRIZ Journal, 9, http://www.triz-journal.com (1998).
19. B. Bohuslav, M. Darrell and J. Pavel, Case Studies in TRIZ: A Novel Heat Exchanger, TRIZ Journal, 12, http://www.triz-journal.com (1999).
20. R. Tan, Innovation Design—TRIZ: Theory of Innovative Problem Solving, China Mechanic Press (in Chinese) (2002).

# OTSM Network of Problems for representing and analysing problem situations with computer support

Nikolai Khomenko and Roland De Guio
*INSA Strasbourg, Graduate School of Science and Technology,*
*Design engineering laboratory (LGECO)*
*24, boulevard de la victoire 67084 Strasbourg cedex FRANCE,*

**Abstract.** This paper presents a method for increasing the level of formalization of the description of a problem situation and obtaining a "big picture" of the problem situation in order to be solved by TRIZ and OTSM instruments. The main output of the method is a list of problems to be turned into contradictions. Elements concerning computer support for this approach are discussed.

**Keywords.** TRIZ, OTSM, inventive problem solving, computer aided innovation.

## 1 Introduction

Before setting out on a complex journey we usually prefer to have a good map of the destination, but when researching in new territories for which no map exists, we have to develop our own map while exploring the area. Analysis of complex non-typical interdisciplinary problem situations could be viewed as a journey into unknown territories. Therefore it is also a good idea to develop maps of the elements of the thought process we pass through. This map will guide us through complex problem situations and help, on the one hand, to collect a set of partial solutions we could use in order to develop satisfactory solutions, and on the other hand, get an holistic view of the links between the problems.

Several problem solving methods and tools based on the idea of a map have been proposed in the past in the area of systems engineering and management like KJ diagram [1], causal loops diagram [2]. Most of these general tools just describe systems and problem relationships and leave the human problem solvers responsible for analyzing and solving the problem. When the cognitive gap between the description of the problem and the description of the solution is too large for the

---

*Please use the following format when citing this chapter:*

Khomenko, N., De Guio, R., 2007, in IFIP International Federation for Information Processing, Volume 250, Trends in Computer Aided Innovation, ed. León-Rovira, N., (Boston: Springer), pp. 77-88.

problem solver additional tools are required. TRIZ [3-6] and OTSM [7-17] [18, 19] theories provide instruments which satisfy this need when the path from problem to solution involves changes in the model of the problem or system. This paper presents a method for increasing the level of formalization of the description of a problem situation and obtaining a "big picture" of the problem situation in order to be solved by TRIZ and OTSM instruments with or without computer support. It is based on the so called "Network of Problems" (NoP) concept and analysis technique which is part of OTSM instruments. It can also be used for problem situation analysis and resolution independently of further use of OTSM problem solving methods. Indeed, as it collects, formalizes and organizes knowledge and information about the problem situation, it can be used at least as any general problem analysis method. Nevertheless, in this paper, we shall focus on the analysis that leads to obtaining the network of contradictions which are to be overcome by the team of professionals and experts in order to solve the addressed problems.

### 1.1 TRIZ in brief

TRIZ is a theory for solving problems during the inventive process. It provides a set of instruments which dramatically decrease the amount of trial-and-errors when a problem situation requires creativity and invention in order to be solved. It was created in the course of an extensive study of the history of engineering systems evolution. As a result of this study three postulates were formulated.

(1) Postulate of objective laws of system evolution. Genrich Altshuller developed a system of 8 laws of engineering system evolution. Based on this system, an instrument for practical needs was developed known as the TRIZ system of standard solutions for inventive problem solving.

(2) The postulate of contradiction states that behind each non typical problem there is a hidden contradiction that should be discovered and overcome in order to solve the problem and reach the next step of engineering system evolution. Based on this postulate, instruments for dealing with contradictions have been developed. The most sophisticated of them, in the framework of Altshuller's work, is ARIZ-85-C, which is a meta-method using most of the basic TRIZ instruments. ARIZ helps to clarify and solve the underlying core contradiction of a problem. ARIZ is also helpful for transforming a problem that seems atypical at the beginning into a typical TRIZ problem description. Then the TRIZ system of standards (typical solutions) could be applied. In case a problem could not be solved by TRIZ typical solutions ARIZ has special tools for dealing with non typical problems.

(3) The postulate of specific situations states that in the course of the problem solving process we should focus on the peculiarities of the problem situation and use available resources of the specific situation in order to study the situation and to construct a satisfactory solution.

ARIZ-85-C integrated all instruments of Classical TRIZ in the united system and based on the theoretical background of Classical TRIZ. In order to develop ARIZ and use general ideas of the axioms in real life situations two main models were also proposed in Classical TRIZ: System operator [6], which is dedicated to describe a problem situation as a whole and its elements in order to simplify the problem solving process; and the TRIZ model of the problem solving process [20].

### 1.2 OTSM in brief

OTSM develops Classical TRIZ ideas further in order to propose instruments to deal with non typical complex interdisciplinary problem situations. The main problem to tackle can be formulated as a question: how to transform all possible problems of invention into one canonical form in order to solve them by a typical problem solving procedure? What should be the canonical form and the procedure for obtaining it?

In the course of our research the driving contradiction that underlies the key questions to be answered by OTSM was indentified: In order to create universal instruments, the rules of these instruments should be as general as possible. But, usually, general rules propose general recommendations that are useless for practice. It means that in order to be useful for real life practice, the rules of the instruments should be specific. But, the more specific the rules are, the narrower the scope of their applications. This contradiction was resolved by using instruments of Classical TRIZ. Then the system of OTSM axioms was developed and axioms of Classical TRIZ were reformulated according to the results of OTSM development. Two models of Classical TRIZ proposed by Altshuller were reviewed and developed further for OTSM purposes. The most important instruments of OTSM are the four main technologies: New Problem technology, Typical Solution technology, Contradiction technology and Problem Flow technology. These technologies are integrated into the Problem Flow Networks approach (PFN) [15] [17, 21] [14] [22]. PFN approach is based on four kinds of networks: Network of Problems; Contradiction Network, Parameter Network (Specific) and Parameter Network [17].

In this paper, we present a method for developing and analyzing the NoP in order to transfer it into a network of contradictions which are to be solved. This method was implemented for the improvement of a power plant. As a result, in March 2006 a patent was obtained. The remaining part of the paper is organized in the following way. First are provided basics about the NoP and an overview of the proposed process. Second the process is described step by step. Third, computer support for this approach is discussed before the conclusion.

## 2 Using the network of problems (general) for analysis

### 2.1 Basic concepts of the network (solutions, partial solutions, edges)

The Network of Problems, which is a high level representation of the problem situation that both gathers and analyzes overall knowledge of the initial situation, can be considered as a semantic network linking several aspects of a many-sided problem situation. The NoP can be considered as an oriented graph the nodes of which represent either problems partial solutions or goals. We define a partial solution as a solution that cannot be generally accepted for one of the following two reasons: (1) the solution solves one problem but produces another one (chain of problems); (2) the solution solves just one or several sub-problems but not the problem situation completely (Sub-domain of the whole problem domain). Goals are specific kinds of problems that will be defined in section 4. The edge meanings are given on Fig. 1.

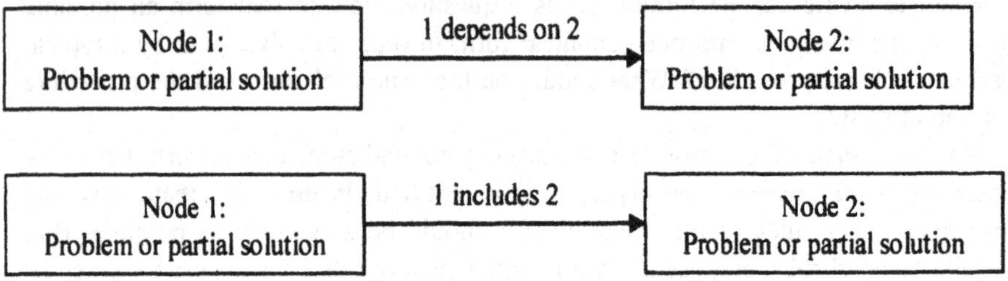


**Fig. 1.** Example of the meaning of the arrows of the Network of Problems

The terminology "super-problem" and "sub-problem" of a given problem are used in reference to the path linking two problems. It is only relevant when the path is not part of a closed loop. A problem A is a super-problem A of a problem B when there exists a pathi from problem A node to problem B node. Inversely in the same situation problem B is a sub problem of problem A. A problem can be both super and sub-problem.

### 2.2 Overview of the process

The first stage of development of the NoP (S1 to S5 in the flowchart Fig. 2) starts by collecting a list of the most painful problems and their potential or partial solutions. Then, relationships between problems and solutions are established. Next, analysis of the actual problem situation is performed according to Classical TRIZ System Operators, a list is compiled of the problems related to certain components of the system and stages of the process the system should perform. As a result of this analysis, a list of problems and partial solution is updated and corrected.

During the second stage (S6 to S12 in the flowchart), the previous list is transformed into a NoP. During this stage, some initial descriptions of problems and potential solutions may be decomposed into sub-problems and solutions. As a result, a map is formed which described the whole problem in a more formalized way. In addition, when the map is developed by a group, it helps to obtain mutual and shared understanding of the problem situation and of the goals to be achieved.

The third stage (S13 to S15) is dedicated to identification and analysis of bottlenecks, which are the most important problems that should be eliminated or bypassed in the course of the problem solving process. Sometimes the problem situation can be resolved at this stage if the participants in the problem solving session have all the necessary skills. Otherwise, the problem solving process goes on by stating contradictions and building OTSM's network of contradictions, which is out of the scope of this paper.

# 3 The analysis procedure step by step (see also flowchart)

## 3.1 The initial list of problems

First, an initial list of the most painful problems should be drawn up without any specific organization. Members of the problem solving team individually or in groups prepare lists of problems which they consider as the most important and painful problems. Each item of the list is a short description of the problem in free form. Then problem description should be clarified with OTSM experts. Eventually participants learn the rules for initial problem representation in the list of problems. If some partial solutions of the problems can be proposed, they have to be collected too.

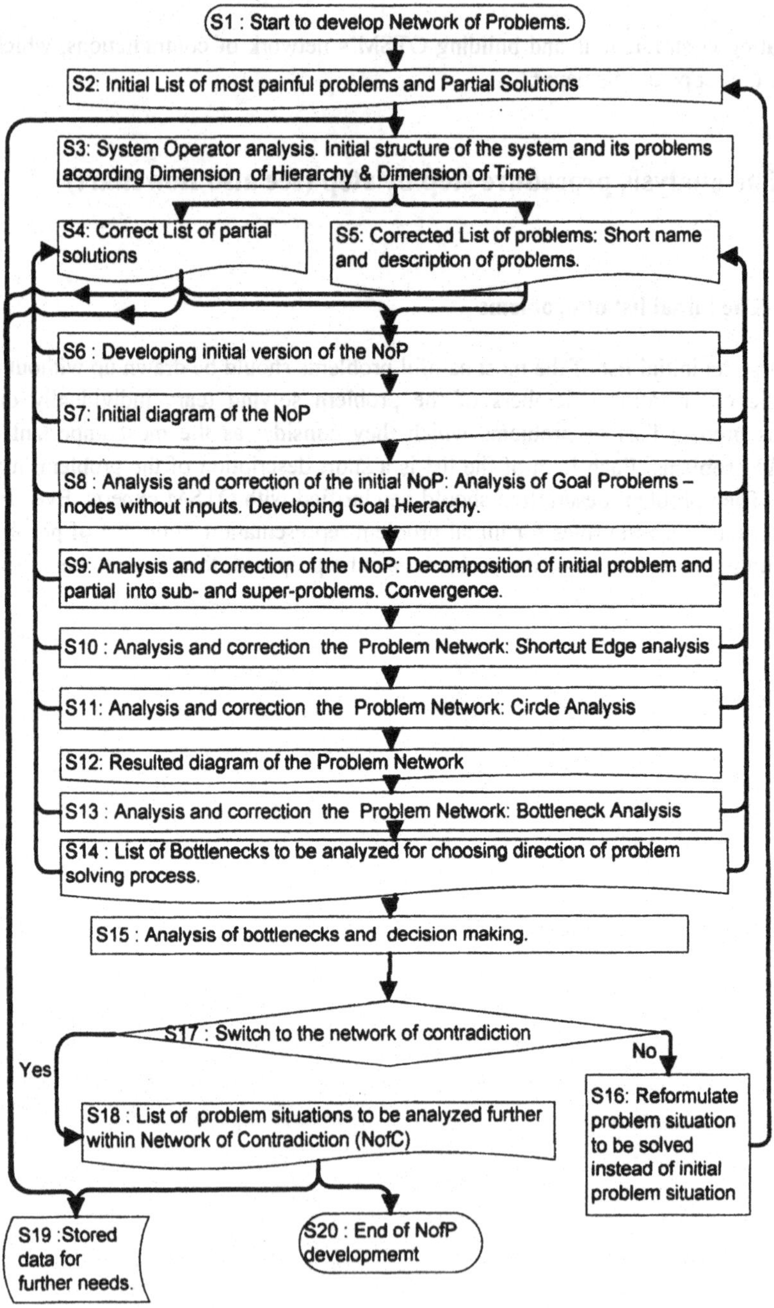


**Fig. 2.** Flow chart of the NoP analysis

### 3.2 System Operator analysis of the problem situationii

It can be observed from the list of problems that each problem belongs to a certain level of system hierarchy (Hierarchy Dimension) and stage of the process that the system should perform according its function (Time Dimension). Thus, the next stage consists in creating a hierarchical schema of the existing or hypothetical system (in case the system is under development) and its technological process (Time Dimension of Classical TRIZ System Operator). This process will reveal additional problems. Each of these problems is relevant to a certain level of the system hierarchy. Therefore groups of problems should be analyzed by level of this hierarchy.

For each additional problem identified during this stage, linked partial solutions should be collected. Any other techniques and methods could also be used in order to update the initial list of problems for the given problem situation. As soon as no new problem appears from Domain Experts, the initial problem network development can start.

### 3.3 Developing the initial version of the Problem Network

The goal of this stage is to establish a relationship between problems and partial solutions in the form of a semantic network (Oriented named graph) like in Fig. 3 below. Problems (Pb) and partial solutions (PS) and, as we will see further, goals are nodes of this network. An arrow linking a problem and a partial solution indicates which problem should be solved or which partial solutions should be implemented in order to solve this problem. When connections between problems and partial solutions are unclear at this stage, but it is likely they are linked, problems and partial solutions are grouped together on the diagram for further clarification of their relationships like in Fig. 4.

Several practical rules for representing the graph, which facilitate further human visual analysis, are used: (1) Arrows should go out of the node box from the bottom side and come into the node box from the top side of the box; (2) problems and partial solution nodes have different colors; (3) the arrows and level of the graph are oriented from the top to the bottom of the page.

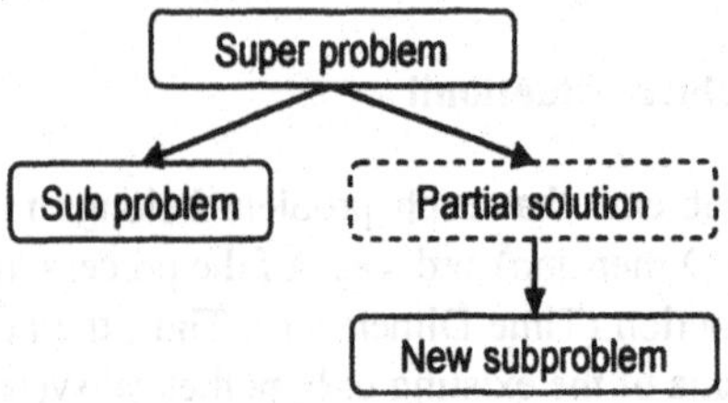


Fig. 3. Arrow direction: from top to down. From super-node to a sub-node

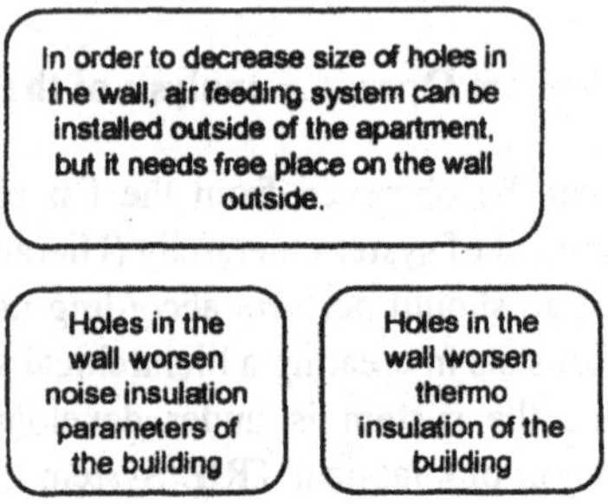


Fig. 4. Sub-set of nodes for further clarification of their relationships with other nodes

The initial network of problems is ready to be analyzed and improved when each problem and sub problem is connected to an arrow.

### 3.4 Goal analysis

At this stage, one should focus on the nodes of the graph that have no incoming arrow. They should be considered as Goals to be achieved. It is necessary to organize them into a system of goals by establishing relationships between goal nodes. Organizing Goal Nodes reveals the set of criteria of good solutions. Good solutions are solutions that help us to solve the top goal problems.

According to the rules of representation given above , these nodes should be located at the top of the graphical representation of the Network of Problems.

### 3.5 Decomposition of problem and solution descriptions

When general criteria concerning the evolution of the required solutions are available, some problems and solutions may be decomposed. Sometimes the problem description contains descriptions of partial solutions as well. Other problem descriptions, which are sub-problems due to implementation of certain partial solutions, may appear. Some partial solutions can be decomposed into several partial solutions or a sub-network of problems and partial solutions.

All of these sub-graphs should be properly integrated into the initial NoP. After a certain amount of practice, decomposition of the problem and solution nodes could be done at earlier stages of the initial NoP development and even whilst gathering problems for the list of initial problems. But at the beginning, it is better to focus on decomposition after organizing the Goal Nodes into a system.

It is important to notice that sometimes decomposition of problem and solution descriptions could lead to a particularly large sub-network of problems. According to the OTSM model of non typical problems, the problem solving process should be presented as a fractal structure. That is why some problems or solutions can be deployed into a sub-network of problems and each network of problems has to be considered as part of a Super-network of problems. For instance the network of problems relevant to a certain project of a company is a sub-network of the whole company Network of Problems. At the same time some sub-problems of the project could be presented as a large network of problems.

### 3.6 Short Cut edge (arrow) analysis.

Sometimes situations are present in the graph where several paths from one problem to another do not have the same number of arrows as in Fig. 6. In this case, it has to be clarified why this shortcut appears. Usually, it shows that either some sub problems are missing or useless. Sometimes we can extract additional information about the initial problem situation and rearrange the whole diagram accordingly.

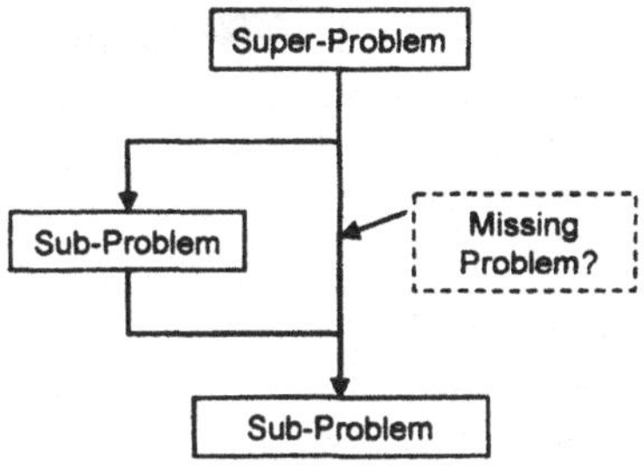


**Fig. 5** Shortcut edge

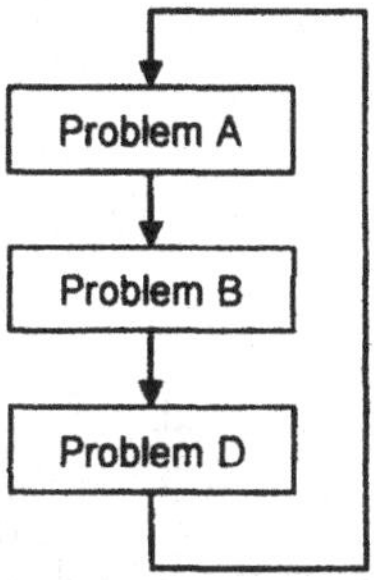


**Fig. 6.** Closed loop of problems

### 3.7 Closed loops analysis

Special attention should also be paid to the loops in sub-graphs like in Fig. 6. They often indicate important hidden contradictions or closed loop situations in the problem situation. Generally, additional information should be gathered and/or new sub-problems and solutions disclosed. As soon as the above mentioned analysis is done and all changes in the list of problems are performed, it can be considered that the development of the initial version of the network is finished and a more precise analysis of the obtained problem situation description can begin (S 13 in the flowchart).

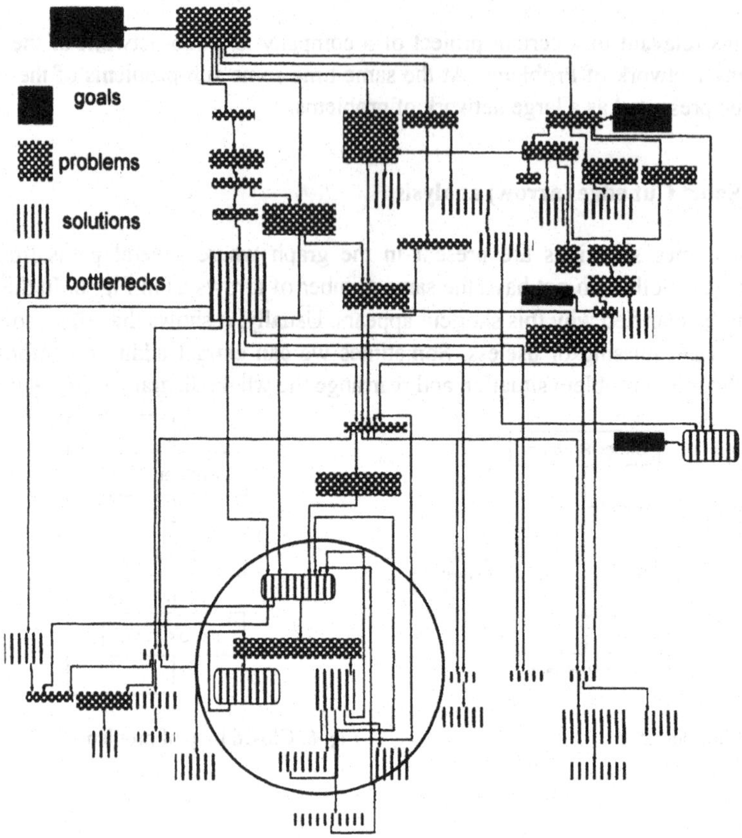


**Fig. 7.** Example of Network of Problem. In the cercle, the bottleneck at the top is involved in two closed loops.

### 3.8 Analysis of the bottleneck problems

A bottleneck problem or solution (see Fig. 7) appears as a node of the network that has several inputs from different problems or solutions. Bottleneck nodes often indicate that at least one hidden contradiction exists in the problem of the bottleneck. Often it is a system of contradictions, which will be developped into a network of contradictions[17]. The point is that the solution of several problems of the project depends on the bottleneck problem. Thus, it is meaningful to focus on these kinds of problems first. A NoP may have several bottlenecks. In this case, it is better to choose those which are closer to the Goal node we are dealing with during the problem solving session. The reason of the previous proposal is that each solution will make the network evolve. When a bottleneck problem is solved, problems below the bottleneck in the hierarchy of problems can disappear with all its sub-problems. Often, bottlenecks nearest to the goal in the top of the network are supposed to have many sub-problems sometimes also bottlenecks. Moreover, the bottom of the hierarchy only shows problems or solutions for which we do not know sub-problems.

After each analysis above, the list of problems should be properly corrected, updated and saved for further needs. As soon as bottleneck and other analyses are performed, it is time to choose a set of problems (bottlenecks, loops etc.) in order to formulate a set of contradictions to be transformed into the network of contradictions.

# 4 Software support for analyzing the Network of problems

Computers can be used and are necessary at several stages of the NoP development. Firstly graphical tools for NoP visualization and layout are necessary. For simple problems, standard tools like Microsoft Visio and several "macros" linking nodes to data bases and problem descriptions are sufficient. Loop, goal and bottleneck identification can be handled visually by the human analyst.

But in the case of large NoPs including several hundreds of sub-problems and partial solutions the capacity for visualization degrades. In this case it could be better to use special software for Graphs or Semantic network analysis. Loops, goals and bottlenecks are standard functions of graph analysis toolboxes. Visual tools for graph hierarchies and clusterings do exist. All these tools can be integrated to build a tool dedicated to this analysis. In practice, we are at the stage of integrating, within a same prototype, tools for NoP development and analysis but also for the next stages of the problem solving process.

# 5 Conclusion

OTSM based NoPs are useful for beginning the complex cross-disciplinary problem solving process. Experts, from various domains, who take part in the sessions for NoP development, notice that this kind of network is helpful for understanding their own problems and finding a shared view of the problem situation. It can be used in order to increase the efficiency of problem solving sessions and to share and organize knowledge and views which are relevant to a given problem situation. That is why we suggest that a process of solving cross-disciplinary problems which is based on the Network of Problems and OTSM could be used as an instrument for knowledge representation and capitalization for further needs of a company or organization.

# Notes

i Sequence of nodes and one outgoing arrow for each node. Path in a graph.
ii System Operator of classical TRIZ is often known as Multi-Screen Schema.

## References


1. Andrew P., S., William B., Rouse, ed. *Handbook of Systems Engineering and Management*. Wiley Series in Systems Engineering. 1999 John Wiley & Sons.
2. Senge, P.M., The Fifth Discipline: The Art & Practice of the Learning Organization, , (Doubleday Currency, , New york, 1990).
3. Altshuller, G.S., To Find an Idea: Introduction to The Theory Of Inventive Problem Solving, (Nauka, Novosibirsk, 1986).
4. Altshuller, G.S., *Algorithm of Invention*, (Moscowskiy Rabochy, Moscow 1969).
5. Altshuller, G.S., *The Innovation Algorithm: TRIZ, systematic innovation, and technical creativity*, (Worchester, Massachusetts: Technical Innovation Center, 1999).
6. Altshuller, G.S., ed. Creativity as an exact science: The Theory of the Solution of Inventive Problems. 1984, Gordon and Breach Science Publishers.
7. Khomenko, N., Modeling of problem situation, in Conference on methodology and techniques of engineering creativity. 1984: USSR, Novosibirsk.
8. Khomenko, N., *Selection of the minimal task*, in *Design research in progress*. 1987, Polish Academy of Science.Institute of philosophy and sociology.
9. Khomenko, N., Tsourikov, V., Contradiction resolution in artificial intelligence software for concept design stage of product developing, in Conference on CAD system for product development. 1988: USSR, Minsk. .
10. Khomenko, N., Software for inventive problem solving training classes, in Conference on Engineering creativity. 1988: USSR, Miass.
11. Khomenko, N., *Contradiction as a system of elementary contradictions*, in Conference on Engineering creativity. USSR, Miass, (Year).
12. Khomenko, N., Using multi dimension space of features for system description, in Conference on Engineering creativity. 1988: USSR, Miass.
13. Khomenko, N., Working Materials for OTSM development: State of Art 1980-1997. 1997.
14. Khomenko, N., *Education Materials for OTSM development: State of Art 1980-1997*. 1999, LG-Electronics Learning Center, Piangteck, South Korea.
15. Khomenko, N., Kucharavy, D., *OTSM problem solving process: Solutions and their classification*, in TRIZ Future 2002 Conference Strasbourg, France, (Year).
16. Khomenko, N., Shenck, E. Kaikov I., OTSM-TRIZ problem network technique: application to the history of German high-speed trains, in TRIZ Future 2006 Conference. 2006: Belgium, Kortijk.
17. Khomenko, N., De Guio, R., Lelait, L., Kaikov, I., A Framework for OTSM-TRIZ Based Computer Support to be used in Complex Problem Management. *International Journal of Computer Applications in Technology* (to be published)
18. Cavallucci, D., Khomenko, N., Morel, C., Towards inventive design through management of contradictions, in 2005 CIRP International Design Seminar. 2005: Shanghai, China.
19. Cavallucci, D., Khomenko, N., From TRIZ to OTSM-TRIZ: Addressing complexity challenges in inventive design. *International Journal of Product Development* 4(1/2): p. 4-21 (2007)
20. Altshuller, G.S., Inventive Problem Solving Process: fundamental steps and mechanisms. . 1975.
21. Khomenko, N., De Guio, R., Cavallucci D.,, Enhancing ECN's abilities to address inventive strategies using OTSM-TRIZ. *International Journal of Collaborative Engineering* (to be published 2007)
22. Khomenko N., Education materials for OTSM, advanced master of innovative design. 2006, INSA de Strasbourg- France.

# A Text-Mining-based Patent Analysis in Product Innovative Process

Liang Yanhong[1], Tan Runhua[2]
Institute of Design for Innovation, Hebei University of Technology, Tianjin, 300130, P.R. China
Email: 1 waterlily00@126.com
2 rhtan@hebut.edu.cn

**Abstract.** Patent documents contain important technical knowledge and research results. They have high quality information to inspire designers in product development. However, they are lengthy and have much noisy results such that it takes a lot of human efforts for analysis. And due to the fact that hidden and unanticipated information plays a dominant role for TRIZ user, it is difficult to discern manually, thus, patent analysis has long been considered useful in product innovative process. Automatic tools for assisting innovators and patent engineers in obtaining useful information from patent documents are in great demand. In TRIZ theory, a product design problem can be considered as one or several Contradictions and Inventive Principles. Text mining could be used to analyze these textual documents and extract useful information from large amount documents quickly and automatically. In this paper, a computer-aided approach for extracting useful information from patent documents according to TRIZ Inventive Principles is proposed.

**Keywords.** patent analysis, TRIZ, Inventive Principles, text mining

## 1 Introduction

A major competitive advantage for any company is the ability of product innovation. With the aid of massive information across the globe, highly complex products are needed to develop to meet the customer needs at a very low cost but in ever-shorter time. Product development is a process, which builds on the basis of knowledge and experience to solve problems. The technical goal of product development is to reduce difference between the initial state and the idealist state of the product, and to look for the best scheme to improve product quality and reliability. Accompany with increasing complexity of products in modern society, the traditional trial is in low efficiency and brainstorming becomes unreliable to product innovation.

*Please use the following format when citing this chapter:*

Yanghong, L., Runhua, T., 2007, in IFIP International Federation for Information Processing, Volume 250, Trends in Computer Aided Innovation, ed. León-Rovira, N., (Boston: Springer), pp. 89-96.

Altshuller, the father of TRIZ, recognized that the place where to look for the basics of innovation and new ideas was not in the brains of inventors, but in the published inventions [1]. In reviewing thousands of patents, Altshuller and his colleague categorized the inventive principles in several retrievable forms, including a contradiction table, 39 Engineering Parameters, 40 Inventive Principles, and 76 Standard Solutions [2]. He provided a systematic process to define and solve any given problem, especially in the field of product development. According to TRIZ, a significant operation of product innovation is to solve design contradictions. When contradictions and Inventive Principles are defined, product is developed by referring to the analogous inventions not only in related fields but also dissimilar problems in other fields that have previously solved the same contradiction.

In recent years, patent analysis [3,4,5] is more highlighted in high-technology management as the process of innovation becomes more complex, the cycle of innovation becomes shorter and the market demand becomes more volatile. Patent documents contain important research results that are valuable for product innovation. However, they are lengthy and have much noisy results such that it takes a lot of human efforts for analysis manually. To obtain useful information, the method by scanning or reading the indexed patent documents from long lists of noisy results, is a rather trivial and time-consuming task that requires a careful manual selection. And the defects of extract information from patent documents, which indexed by standard keyword-based search methods, will ignore relevant solvable schemes and enlightenment in other fields. In addition to the huge requirement of manpower and time, the rapid increase of the number and application of patent documents, thus, there is a need to find a way to get ride of tradeoffs and obtain useful and precise patent documents quickly. One method to solve this difficult problem is data mining. Data mining is an automated scheme to extract useful information from large databases. As to patent documents are nearly unstructured texts, text mining, like data mining or knowledge discovery, which specialized for full-text patent analysis, is applied to derive information.

Thus, automatic tools of text mining in patent analysis for assisting innovators or patent engineers are in great demand.

Further elaboration on text mining and patent classification would be provided in section 2. Following that, elaboration on the background of text mining for patent analysis would be provided in the rest of paper. Following that, a text-mining approach that helps extract and analyze useful information automatically from online network-patent documents is discussed. Thus, automatic tools and softwares for assisting TRIZ users and patent engineers to select useful patent documents from large amount of patents are given. Further, some of the difficulties in extracting information not only concise but also not loss are also presented. Finally, future research work is discussed. Our focus, however, would be largely directed towards the methodology to extract textual components from patent documents.

# 2 Theoretical background

## 2.1 Patent classification and characteristics of patent document

Patent classification schemes are used to organize and index the technical content of patent specifications so that specifications on a specific topic or in a given area of technology can be identified easily and accurately. Before their publication, patent documents are given one or more classification codes based on their textual contents for topic-based analysis and retrieval. Many patent classification schemes, such as IPC (International Patent Classification), US Classification and British Classification, have been developed. However, the classification schemes used by these researchers are based on the application fields involved in the inventions, For example, the IPC divides patent technology into 8 key areas:

A: Human Necessities
B: Performing Operations, Transporting
C: Chemistry, Metallurgy
D: Textiles, Paper
E: Fixed Constructions
F: Mechanical Engineering, Lighting, Heating, Weapons
G: Physics
H: Electricity

Patent documents are divided into different areas according to technology fields is helpful to search the prior art for traditional inventors. However, it is inadequate for TRIZ users since TRIZ users are interested in previous patents that have solved the same Contradiction and used the same Inventive Principles, which may come from different fields [6,7]. Patent documents, which are classified according to Inventive Principles combined with Contradiction, will provide a broader view for TRIZ users and TRIZ software developers, by helping them find possible inspiration from a field that may be totally different from theirs.

A patent document contains dozens of items for analysis. Some are structured, that is to say they are uniform in semantics and in format such as patent number, filing date, or assignees; some are unstructured, that is to say they are free texts of various lengths and contents, such as title, claims, abstract, or descriptions of the invention. The description of the invention can be further segmented into field of the invention, background, summary, and detailed description, although some patents may not have all these segments. Patent analyses based on structure information have been the major approaches in practice and in the literature for years [8,9,10]. These structured data can be analyzed by data mining techniques or well-established database management tools such as OLAP (On-Line Analytical Processing) modules. But the most rest of patent document is made of unstructured text, based on this, there has been an interest in applying text mining techniques to assist the task of patent analysis. Based on Text mining people do not have to understand text in order to extract useful information from it.

## 2.2 Text mining

Data mining, also known as knowledge discovery in a database, is a recent development for accessing and extracting information in a database [11,12]. In short, data mining applies machine-learning and statistical analysis techniques for the automatic discovery of patterns in a database. Most efforts in data mining, however, have been made to extract information from a structured database and the utility of data mining is yet limited in handling huge amounts of unstructured textual documents.

As a remedy, text mining [13,14] is a rather new technique that has been proposed to perform knowledge discovery from collections of unstructured text. Like data mining or knowledge discovery, text mining is often regarded as a process to find implicit, previously unknown, and potentially useful regularities in large textual datasets. In briefly speaking, text mining puts a set of labels on each document, and discovery operations are performed on the labels. The usual practice is to put labels to words in the document. Then, the document in text format can be featured by keywords and clue words that are extracted through text mining algorithm. Since it is suitable for drawing valuable information from large volumes of unstructured text, text mining has been widely adopted to explore the complex relationship among patent documents. With the application of text mining an effective means can be provided for content searches in the textual fields of patent documents.

## 3 A text-mining-based methodology to patent analysis

Based on the theoretical background introduced above, a general methodology is suggested to analyze patent documents. The overall process of conducting text-mining-based patent analysis goes through several steps. First of all, text collection and text preprocessing are the preliminary step. The interested patent area is selected and related patent documents are collected in electronic text format. Second, raw patent documents are transformed into structured data. Since the original documents are expressed in natural language format, they must be transformed into structured data in order to be analyzed and utilized. Text mining that extracts keywords and clue words from patent document is used to this end. Fig.1 depicts the overall process of text-mining-based patent analysis [15]. In relation to patent analysis, text mining is used as a data processing and information-extracting tool. Since the original patent documents are expressed in natural language format, it is necessary to transform raw data into structured data. Then, the process of keyword and clue word extraction is applied to measure similarity between patents.

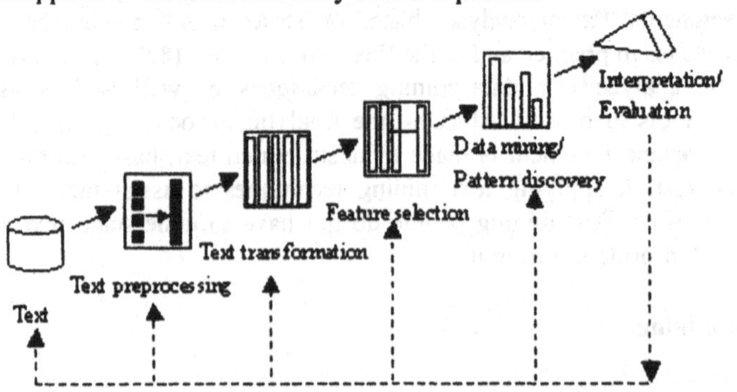


**Fig.1.** Text mining process of patent analysis

### 3.1 Text preprocessing

The patent documents in our experiment from USPTO(United States Patent and Trademark Office: www.uspto.gov) are in HTML format and contains title, abstract, claims, and description. When manually analyze the documents, we found that usually the abstracts and descriptions provided enough semantic information to determine TRIZ Inventive Principles that the patents used. Therefore the abstracts and descriptions are extracted from the full text and parsed into independent sentence. As pointed out above, raw documents need to be processed because they are unstructured in format. The description, the main body of the detailed content, often have sub-sections with titles in uppercase, such as FIELD OF THE INVENTION, BACKGROUND, SUMMARY OF THE INVENTION, and DETAILED DESCRIPTION OF THE PREFERRED EMBODIMENT. Although some patents may have more or fewer such sub-sections or have slightly different title names, most patents do follow this style. Thus a regular expression matcher is devised to extract each of these segments. It is implemented using regular expressions in Perl. The method takes advantage of the rule that each sub-section's title is in a single line paragraph separated by two HTML tags: "<BR><BR>". After splitting the paragraphs based on these tags, a set of Perl expressions: (/Abstract/i, /Claims/i, /FIELD/i, /BACKGROUND/i, /SUMMARY/I, /DESCRIPTION|EMBODIMENT/i) are used to match the patent segments.

### 3.2 Text transformation and feature selection

The document is segmented at word level as the smallest unit. The document is first split into a series of words. Each document is made of bags of words. Adjectives, adverbs, nouns and multi-word are extracted from the document. Word frequency (term frequency) and inverse document frequency are two parameters used in filtering terms. Low TF and DF terms are often removed from the indexing of patent documents. After removing stop words [16] in each document, word stemming [17] is performed. To better match concepts among terms, words are stemmed based on Porter's algorithm [18]. It contains keywords, title words, and clue words. Here the keywords are those maximally repeated words that can be calculated by a fast key term extraction algorithm [19]. The algorithm works with the help of a stop word list alone. By repeatedly merging back nearby words based on three simple merging, dropping, and accepting rules, Maximally repeated strings in the text are thus extracted as keyword candidates. The algorithm is shown in fig.2. The title words are those non-stop words that occur in the title of a patent document. As to the clue words, they are a list of about 25 special words that reveal the intent, functions, purposes, or improvements of the patent. These words are prepared by several patent analysts based on their experiences and are listed as table 1.

```
For each patent document
{
    1.1 Convert the text into a LIST of words
    1.2 Do
    {
      2.1 Set MergeList to empty
      2.2 Put a separator to the end of LIST as a sentinel and set the occurring frequency of the separator to 0
      2.3 For I from 1 to NumOf(LIST) - 1 step 1
      {
          3.1 If LIST[ I ] is the separator, then
                Go to Label 2.3.
          3.2 If Freq(LIST[ I ]) > threshold and Freq(LIST[ I+1 ]) > threshold, then
                Merge LIST[ I ] and LIST[ I +1] into Z
                Put Z to the end of MergeList
              Else
                If Freq(LIST[ I ]) > threshold and LIST[ I ] did not merge with LIST[ I - 1], then
                    Save LIST[ I ] in FinalList.
                If the last element of MergeList is not the separator, then
                    Put the separator to the end of MergeList.
      }
      2.4 Set LIST to MergeList.
    }while NumOf(LIST) < 2
    1.3 Filter terms in FinalList based on some criteria
}
```

**Fig.2.** The keyword extraction algorithm

**Table 1.** the clue words for the BACKGROUND segment

| Advantage | Difficult | Improved | Overhead | Shorten |
|---|---|---|---|---|
| Avoid | Effectiveness | Increase | Performance | Simplify |
| Cost | Efficiency | Issue | Problem | Suffer |
| Costly | Goal | Limit | Reduced | Superior |
| Decrease | Important | Needed | Resolve | Weakness |

### 3.3 Pattern discovery

Whenever the designer faces a problem in product development, he would search for inventive experience and information to enlighten his thinking. However, this search is not an easy one to TRIZ user. A keyword search is helpful to the innovator but sometimes the useful patents maybe be neglected since the same Inventive Principles to solve the problem come from different fields. Therefore there is a need to accurately return records that might exhibit similar problems and causes to innovators. Under these circumstances, clustering might be a possible tool that could be employed. Records could be initially placed into groups using clustering algorithm [20] and multi-Naïve Bayes algorithm [21,22,23]. Then, as TRIZ user faces a new problem, that is to say, when an Inventive Principle is given, records that exhibit similar problems to the groups could be returned.

## 5 Conclusion and future work

This paper describes characteristics of patent document and text mining technique. A methodology based text mining that could be used to analyze patent document for TRIZ user is presented. It points out clustering algorithm and multi-Naïve Bayes algorithm to extract useful information from patent document, which is helpful in product innovative process for TRIZ users. However, more work is required and some difficulties exist. Firstly, to put the methodology into practice, an automated text mining system will be developed using Perl and WEKA software [24]. Secondly, in addition to TRIZ Incentive Principles, Contradictions would also be taken into consideration. The difficulty is that the textual dataset of patent is not a numerical dataset. Users in different circumstance, or with different needs, knowledge of linguistic habits will describe the same information using different terms. What is more, sometimes the same word could have more than one distinct meaning. These add to difficulties of understanding the texts.

## 6 Acknowledgement

The research is supported in part by the Chinese national 863 planning project under Grant Number 2006AA04Z109. No part of this paper represents the views and opinions of any of the sponsors mentioned above.

## 7 References

1. R.H. Tan, *Theory of Inventive Problem Solving: TRIZ* (Science Press, China, 2004).
2. http://www.oxfordcreativity.co.uk/ (April 10, 2007)
3. V.W Soo, S.Y Lin, S.Y Yang, S.N Lin, S.L Cheng, A cooperative multi-agent platform for invention based on patent document analysis and ontology, *Expert Systems with Applications 31* (2006), pp766-775
4. B. Yoon, Y. Park, A text-mining-based patent network: Analytical tool for high-technology trend, *Journal of High Technology Management Research 15*(2004), pp. 37-50.
5. G. Fischer, N. Lalyre, Analysis and visualization with host-based software-The features of STN AnaVist, *World Patent Information 28*(2006), pp. 312-318.
6. H.T. Loh, C. He, L.X. Shen, Automatic classification of patent document for TRIZ users, *World Patent Information 28*(2006), pp. 6-13.
7. C. He, H.T. Loh, Grouping of TRIZ Inventive Principles to facilitate automatic patent classification, *Expert System with Applications* (2006), pp1-8
8. D. Archibugi, M Pianta, Measuring technological change through patents and innovation survey, *Technovation*, 16(9) (1996) , pp.451-468.
9. B.D.Carrax, C. Hout, A new methodology for systematic exploitation of technology databases, *Information Processing & Management*, 30(3) (1994), pp407 - 418.

10. K.K Lai, S.J. Wu, Using the patent co-citation approach to establish a new patent classification system, *Information Processing & Management*, 41(2) (2005), pp313 - 330.
11. U. Fayyad, G. Piatetsky-Shapiro, P. Smyth, From data mining to knowledge discovery In Databases, *American Association for Artificial Intelligence* (1996), pp37-56.
12. G. Piatetsky-Shapiro, U. Fayyad, P. Smyth, Advances in knowledge discovery and data mining, *AAAI/MIT Press* (1996), pp1–33.
13. I.H. Witten. Text mining, *Practical handbook of Internet computing* (CRC Press, Boca Raton, 2004).
14. Witten I.H. Bray. Z, Mahoui. M, and Teahan .W, Text mining: A new frontier for lossless compression. *Proceedings of the Data Compression Conference*, Snowbird, UT.Los Alamitos, CA:IEEE Computer Society Press (1999), pp.198-207.
15. A. Wasilewska, cse634-Data Mining: Text Mining (June 5, 2007); http://www.cs.sunysb.edu/~cse634/presentations/
16. Onix Text Retrieval Toolkit (June 5, 2007); http://www.lextek.com/manuals/onix/stopwords1.html.
17. Porter Stemming Algorithm (March 5,2007); http://www.tartarus.org/~martin/PorterStemmer/.
18. M.F. Porter, An algorithm for suffix stripping, *Program* (1980), 14(3), pp130-137.
19. Y.H. Tseng, C.J. Lin, Y.I Lin, Text mining techniques for patent analysis, *Information Processing and Management* (2007)
20. A. Gatt, Structuring Knowledge for Reference Generation: A Clustering Algorithm (March 2, 2007); http://www.csd.abdn.ac.uk/~agatt/home/pubs/
21. Y. Yang, X. Liu, A re-examination of text categorization methods, *SIGIR99* (1999).
22. I.H. Witten, E. Frank, Data Minging: *Pritical Machine Learning Tools and Techniques, Second Edition* (Elsevier Inc, Singapore, 2005).
23. A. McCallum, K. Nigam, A comparison of event models for Naïve Bayes text classification, *Proceedings of the AAAI-98 Workshop on Learning for Text Categoration, Madison, WI.Menlo Park,* CA:AAAI Press(1998), pp41-48.
24. WEKA (September 2, 2006); http://www.cs.waikato.ac.nz/ml/weak/.

# Technology Innovation of Product Using CAI System Based on TRIZ

Zhang Jianhui, Yang Bojun, Tian Yumei and Tan Runhua
Institute of Design for Innovation, Hebei University of Technology
300130, Tianjin, P.R.China,
zjh2031@sina.com
home page: http://www.triz.com.cn

**Abstract.** Technology innovation and creative problem solving are necessary for long-term enterprise survival. Making strategic decisions for product development is a key step in the technical innovation process. Computer-aided innovation (CAI) systems based on TRIZ (Theory of inventive problem solving) support strategic ideas generation of innovation. TRIZ methods are applied to assess the maturity of a technical system and to forecast future technological R&D plans. A general process for technology innovation of product supported by CAI system is developed. A case study is presented to show how technology maturity mapping and patterns and lines of evolution are used to predict the future development of butterfly valve's sealing technology.

## 1 Introduction

Making strategic decisions for product development is one of the toughest jobs that managers of Research and Development have in an organization. Deciding between optimizing existing technologies and developing new core technology is one of them [1]. TRIZ (Theory of inventive problem solving) is emerging as a powerful scientific tool that is applied to assess the maturity of a technical system and helps decision-makers to make strategic forecasting decisions.

Every product is a technical system based on its core technology, and the technical system evolves over time, which process is a macro-level methodology aimed at the maturity of an existing technology. What evolve are characteristic indices of the technical systems and its improvement can be shown by S-curve on the time-scale. The stages of the S-curve are infancy, growth, maturity and decline [2,3]. The position of current product on its evolutionary S-curve is called the technology maturity of product (TM) and the process of identifying TM is called the product technology maturity mapping (TMM) [1,4].

*Please use the following format when citing this chapter:*

Jianhui, Z., Bojun, Y., Yumei, T., Runhua, T., 2007, in IFIP International Federation for Information Processing, Volume 250, Trends in Computer Aided Innovation, ed. León-Rovira, N., (Boston: Springer), pp. 97-106.

Understanding the TM allows strategic decisions to be made concerning the optimization or innovation related to product development strategies. Once the TM of product is determined, patterns and lines of technological evolution belong to technological forecasting tools of TRIZ can be used to find some technology opportunities and to forecast future technological R&D plans. The Computer-aided innovation (CAI) systems based on TRIZ are practical tools to support this process, which can accelerate the technology innovation process of product.

This paper introduces a new method for TMM and give elements to guide the optimization or innovation decision involved in product development strategies. Technology forecast tools of TRIZ are applied to help decision makers to determine the best way forward. A general process for technology innovation of product supported by CAI system is formed.

## 2 Technology Maturity Mapping

Technology advancement is a principal impetus in economic development. Foreseeing technological advancement that will shape the future is of immense importance for many organizations, since they can be deeply influenced by emerging innovations [5]. Assessment of an organization's current technology should drive the direction of the R&D planning process. Ellen Domb [8] Suggest that "people tend to do an initial assessment of their product maturity base on their emotional state." An assessing technology, based upon a systematic logical analysis, should be introduced.

Altshuller found out that technical system evolution as a function of time was related to four primary S-curve descriptors: the number of inventions in the field, the level of those inventions, the performance of the technology, and the profitability of the technology [2]. Data pertaining to these descriptors can indicate the position of the technology on the main biological S-curve, so the maturity of technical system can be assessment [1,4].

But to some products the performance index and profitability index could not be expressed exactly by a single parameter that can be gotten directly, and sometimes the data could not be gotten [6]. For example, different industries each have different curves for profitability vs. time. Darrell Mann [7] mainly examined the quantity distribution of two kinds of patent, which are number of cost reduction related patents and symptom curing related patents, to refine the maturity of a system. The number of such inventions tends to increase as the system matures.

Integrating the fruits of Altshuller and Darrell Mann on TMM, Zhang [6] uses patent level, patent number and number of symptom curing patent (SCP) as criteria to map the technology maturity of a product. The standard curves are shown in figure 1. Patent number represents the active degree of technical innovations of the product. The patent level reflects the level of innovations. Altshuller divided all patents into five levels to measure the importance of the patents [2,3]. Symptom curing patent means patents which focus on curing problems which emerge as a result of earlier inventions by introducing implement technology, structure or parts rather than search for substitute technology without relative limitation [7]. And the change of the quantity distribution shows the shift of the improvement keystone.

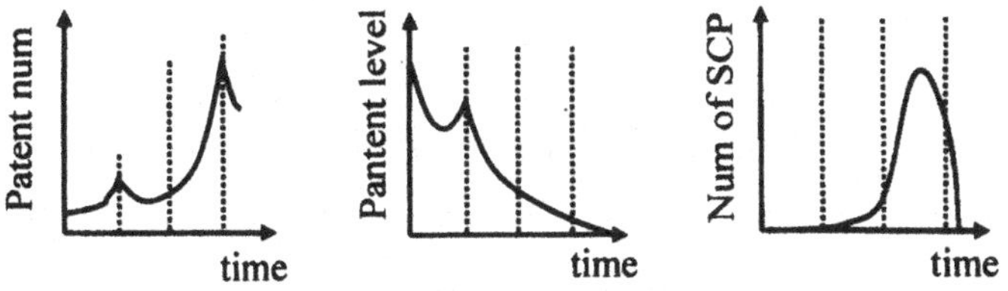


**Fig. 1.** Product Patent Attribute Curves.

The three curves share common data source, and each shows a patent attribute of product respectively [6]. Each of the Product Patent Attribute Curves (PPAC) can be constructed by collecting patent data that relevant to the system under study. The shapes of each of the curves are compared with the shapes of the classical curves, and then a composite analysis of the three curves reveals a data-driven assessment of the maturity of the technological system.

## 3 Optimization or Innovation Dilemma Solving Using CAI

Once the TM of a company's product has been assessed, the future R&D direction can be decided by decision-makers. A strategic decision for company is the dilemma of the optimization versus innovation. Should further investments be made to optimize the technology around the core technology? Or should investments be made to develop a new core technology to replace the existing core technology [1]?

To generate strategic ideas successfully and reduce uncertainties effectively are essential to the future R&D direction. The point of view of this study is that the patterns and lines of evolution belong to technological forecasting tool of TRIZ can be used to generate ideas in a systematic manner with reducing uncertainties. In addition, most of the tools of TRIZ are include in CAI systems. So strategic decision generation supported by CAI systems is possible.

Generally, the patents of the existing product should be gotten first and TMM then should be made. If the core technology of the product under study is in the mature or decline stage, actions of innovation of the core technology should be taken firstly. If the core technology is in the infancy or growth stage, actions of optimization of the core technology should be taken firstly. Once the TM of a product is determined, the patterns and lines of technological evolutions can be applied to forecast future technological development.

The development CAI systems based on TRIZ has made TRIZ more applicable and practical. A few CAI systems, such as Goldfire Innovator of Invention Machine (USA), IWB of Ideation International (USA), InventionTool (China) have been developed and applied by industries. They include data-base for technological evolution, inventive principles, effects, etc. And the InventionTool has the TMM function module. The application of these systems makes the designers to find problem solutions more easily in a short time.

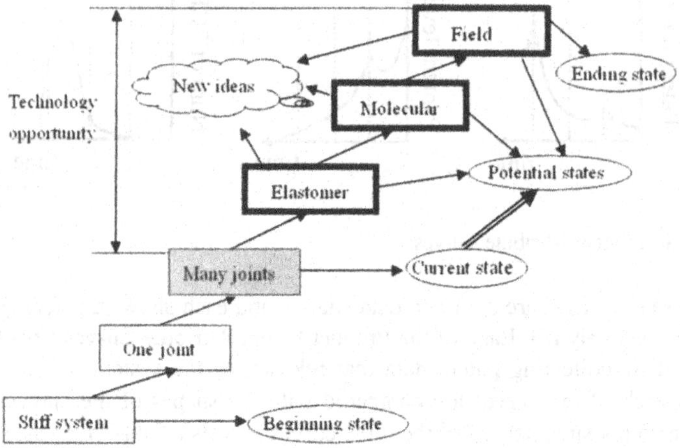


**Fig. 2.** A line of technological systems evolution

# 4 Technology Opportunity Identifying

Patterns and lines of technological evolution belong to technological forecasting of TRIZ [8], which is a proactive approach to forecasting developed during the 1970s

and now is also in development. Technological Forecasting reports the probabilities of certain design parameters falling within particular confidence intervals at some future time. Critical design advances for future products and processes can be identified by applying the patterns and lines. With these advances, the field of parameters is narrowed and a tighter range for confidence levels is defined.

The patterns are very helpful for technology forecasting since they identify the most effective directions for the system's development. A pattern of evolution delineates a general direction for further system transformation but says nothing about the details of this transformation. The lines of technological evolution under each pattern describe more specifically the stages of the systems development and therefore provide even greater predicting power.

Figure 2 is a line under pattern: Evolution Toward Increased Dynamism and Controllability. There six states shown in this line which are stiff system, one joint, many joints, elastomer, molecular and field. The beginning state for this line is stiff system and the ending state is field. If the technology of the product under study is in the state of "many joint", the future states of technology evolution may be elastomer, molecular, or field. Every future state may imply one or a few ideas for the future development of the technology. The distance between the ending state and the current state in that line is a technology opportunity or a technology gap. From the line a technology opportunity can be found and several new ideas can also be generated [9].

Technology opportunities for the existing product under study can be found by searching different patterns of evolution. First, a pattern is selected and a line under this pattern is also selected. Then, current state for the product is determined from the line selected. By this searching method one or a few technology opportunities can be identified.

Every technology opportunity includes one or several potential states, which are upper states from the current to the ending state. Every potential state implies one or a few new ideas of potential technologies to be used for the evolution of current technology in the future development.

## 5 A general process for technology innovation of product supported by CAI system

The technological systems evolve over time with the functional efficiencies of key characteristics of the technical system increase dramatically. Each function is associated with a certain part of the technical system (product or process), and the various forms of technical innovations indicate the evolution of functions. The key patents about a certain product are the track record of core technical innovation of the product and can represent the history of product evolution. So the patent information of the product can be treated as the main source of product evolution analysis. A general process of new idea generation for technology innovation supported by CAI includes the following six steps:

Step 1: Product patent retrieval and filtration

To an enterprise the first work is to define its local or international competitor, to select patent databases in competitive range, and to use the key word to retrieve all patent relevant to the product. In addition, those irrelevant patents to the object evolution in nature should be excluded. For example, the patent about product's appearance is irrelevant to the technological evolution of product, and can be excluded.

Step 2: Analyze the history of product evolution and assess its TM

The analysis of key patent information can provide with valuable technical and business information regarding target or key technology of the product. At the same time, patent level, patent number and number of SCP can be available, so the PPAC can be constructed and the TM of the product can be mapped. InventionTool3.0 (China), which is a CAI system, supports this step.

Step 3: Decision between optimization and innovation

If the core technology of the product under study is in the mature or decline stage, innovation of the core technology is preferred. If the core technology is in the infancy or growth stage, optimization of the core technology is preferred.

Step 4: Select relevant patterns and lines of evolution in CAI systems to analyze the product and make strategic decision

Apply the patterns and lines that matched the evolution of product under study to analyze the product and making a connection. So the current state and potential states can be determined, and the technology opportunity can be identified to forecast future technological development.

Most of the tools of TRIZ are include in CAI systems. And the CAI systems, such as Goldfire Innovator of Invention Machine (USA) and InventionTool 3.0 (China) can support this step.

Step 5: Generate strategic new ideas

Under analyzing the technology opportunity that founded in the relevant line, some strategic new ideas for future R&D direction of product can be generated.

Step 6: Evaluate new ideas

Decide if it is feasible to support possible future goals using the new ideas, and if it is, then select the development to be implemented, otherwise, ignore the ideas.

## 6 Case Study

Butterfly valve is a controlling valve used in large-size pipe, which has function of opening or closing pipe, regulating flux and check. It initially comes from the baffle of chimney or flue that regulate the scavenge capacity. And today it has been used abroad in chemicals, metallurgy, waste treatment, and so on. A butterfly valve has several characteristic indices such as integral leakage, maximum working press, times of open and close, maximum rotational moment. And yet one index improvement can make another decline. Then finding an integral characteristic index is not duck soap.

Step 1: Product patent retrieval and filtration

It is difficult to know when the first butterfly was designed and developed in China. The key word "butterfly valve" is used to retrieve relevant patents from Chinese patent base. 688 patents are retrieved and some of them were eliminated and at last about 600 patents from 1985 to 2005 are selected as resources to carry out this study.

Step 2: Analyze the history of product evolution and assess its TM

According to study tends of the technology evolution for the main structure of the butterfly valves, it is divided into main three modules: the controlling module (includes drive and transmission section), executing module (as shown in figure 3, includes valve disc, valve shaft, valve seat, and seal ring) and valve body. The evolution of the executing module shows the core technology changes of the seal structure of this kind of valve.

The CAI system, InventionTool3.0 is applied to carry out the TMM. After determining the level, totaling year after year, proportion of SCP, these information are input into the TMM functional module of InventionTool. Then conic and cubic are selected to fit curves, which are shown in figure 4 with the classical curves. In the last five years the proportion of SCP to total was about 70%. By analysis of the Product Patent Attribute Curves in figure 4, a conclusion is made that the butterfly valve of the country is in its maturity stage.

Step 3: Decision between optimization and innovation

By the conclusion in step 2, the butterfly valve technology can look forward to a new core technology. Though, optimization of the auxiliary, secondary or harmful functions can keep on gain profit but it will not allow the valve firm to stay competitive for the long run. Decision should be make now to insure a long-term profit. In recently years, much demand has arisen for new substitute technology and it is likely that this trend will continue. So, further investments should be made to develop the new technology around the core technology.

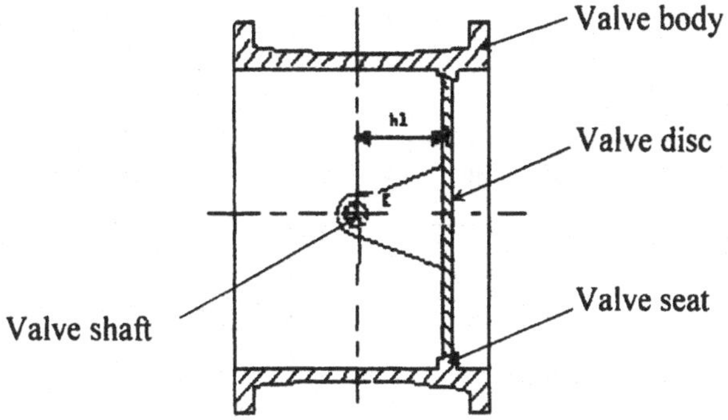


**Fig. 3.** The main structure of butterfly valve

Step 4: Select relevant patterns and lines of evolution in CAI systems to analyze the product and make strategic decision

On the basis of the above analysis, patterns and lines of evolution should be applied to develop new core technology. Almost half of butterfly valve patents since 1985 have related to sealed issues, in order to improve the integral leakage index. Such as, a main trend of the technology evolution for the main structure of the valves is that the valve shaft and the valve disc in early valve are connected immovably (such as Chinese patent No. GG86208284). Later, they are connected jointly by one or more joints (such as Chinese patent No. ZL96241794.7 and No. 87208019U). Now, they are connected flexibly in some valves (such as Chinese patent application No. 94111763.4, the valve disc and valve shaft connected by a spring mechanism). So, evolution line of 'increased dynamics', which matched to this trend, was selected to find the technology opportunities in relation to the seal technology in the evolution of the butterfly valve. InventionTool 3.0 (China) is applied in this step.

Figure 5 shows the evolution line of increased dynamics:

stiff system→a joint→many joints→elastic→molecular (liquid, gas)→field

The current state for evolution of the main structures of the valve is in the "elastic", which matches the flexibility structure in some valves, and the states of "molecular" and "field" are potential states. The distance between the "field" state and the "elastic" state in the line exists the 'technology opportunities'.

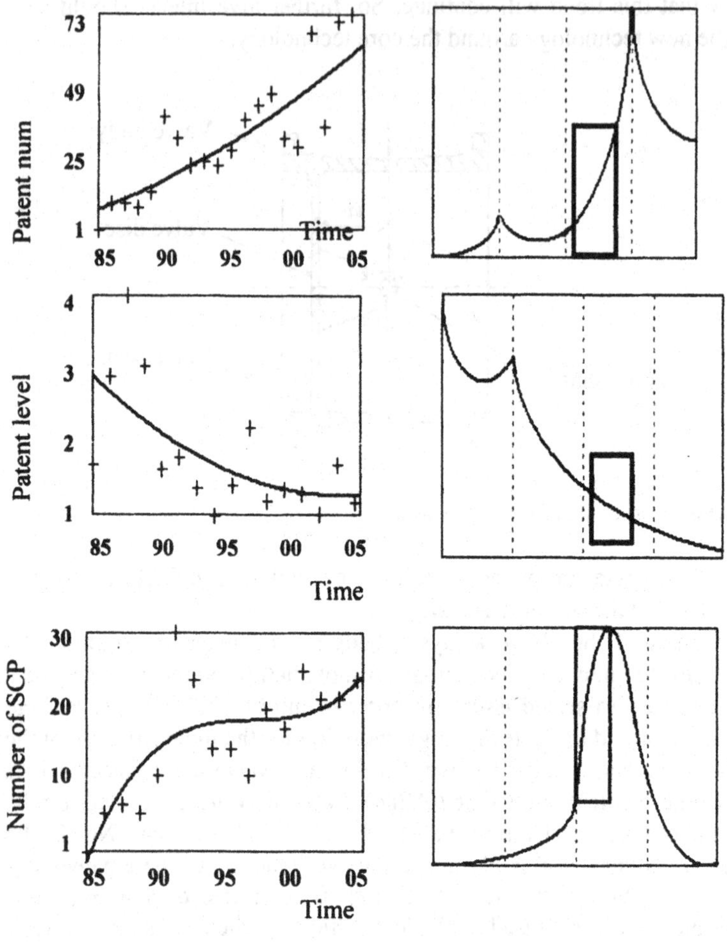


**Fig. 4.** The Product Patent Attribute Curves of butterfly valve versus classical curves

Step 5: Generate strategic new ideas

The key technique of the valve is to compensate for the wear of the seal ring and the valve disc end in order to reduce leakage. The action of the compensation is to make the valve disc to move forward in a small displacement after the rotation of the valve disc when closing the valve, or to make the valve disc to move backward in a small displacement before the rotation of the valve disc when opening the valve.

The application of the compressibility of oil or gas can implement the compensation function from the first potential state "molecular", the second potential state is also applicable. For example, a permanent magnet can implement the function of the compensation. So, some new ideas are generated:

Innovative ideas: Apply compressibility of liquid or gas, or a permanent magnet to implement the function of the seal ring compensation.

Step 6: Evaluate new ideas

These innovative ideas should be possible because there are successful applications in industry or life. And they are generated from the 'technology opportunity' that other successful industries have already made, so the uncertainty of the new ideas is reduced. The engineers of the butterfly valve industry should also make evaluation for the ideas and implement them by the conditions of management, design and manufacture capacity of their firm, they should make an adaptive and specific design for the existing products that need to be improved. Accordingly, the pace of the products innovation is accelerated.

## 7 Conclusions

To generate innovative ideas for product development is a key step in the technical innovation process. CAI systems based on TRIZ are applicable to support this process. An integrated method is applied to assess the maturity of a technical system, and the result of TMM can guide the designer to decide between the optimization and innovation of the core technology. Then patterns and lines of technology evolution can be applied to find some technology opportunities of products and to forecast future technological R&D plans.

A general process for technology innovation of product supported by CAI system is developed, with the aim of making the technology innovation more easily and practical. Technology innovation of the seal structure of butterfly valve is carried out and demonstrates the process step by step. The other patterns and lines of evolution can be applied to generate additional new ideas to provide a picture of the overall possible technological developments in butterfly valve technology, and to provide the decision-makers a striving direction for the new engineering products and technological innovation.

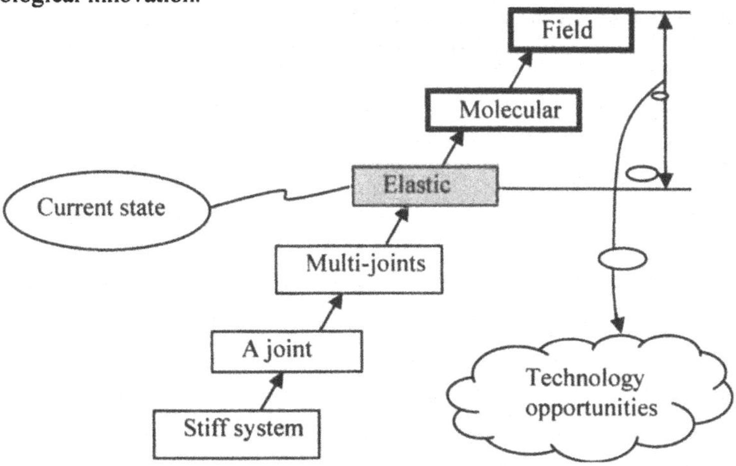


**Fig. 5.** 'Increasing the degree of system dynamics' evolution line.

## 8 Acknowledgments

This research is supported by Natural Science Foundation of China under Grant Numbers 50675059, Chinese national 863 planning project under Grant Number 2006AA04Z109 and Education Office of Hebei Province of China under Grant Number 20062015. Authors are grateful to their fund for this work.


## 9 References


1. Severine Gahide, Application of TRIZ to Technology Forecasting Case Study: Yarn Spinning Technology, TRIZ Journal, July, 2000.
2. Altshuller G.S., Creativity As An Exact Science: The Theory of the Solution of Inventive Problems, The United Kingdom: Gordon and Breach Science Publishers Inc, London, 1984, pp. 205-216.
3. Tan Runhua, Innovation Design——TRIZ：the Theory of Inventive Problem Solving, China Machine PressBeijing, 2002, pp. 58-60. (in Chinese).
4. Michael S. Slocum, Technology Maturity Using S-curve Descriptors, TRIZ Journal, April, 1999.
5. Fey V R, Rivin E I., Guided Technology Evolution, http: //www.trizgroup.com/article2.html, July, 2001.
6. Zhang Huangao, Tan Runhua, and Zhao Wenyan, The technology maturity of product and its mapping. ASME Proceedings of IMECE'04, IMECE2004-59260. Anaheim, California, USA, 2004.
7. Darrell Mann, Using S-Curves and Trends of Evolution in R&D Strategy Planning, TRIZ Journal, July,1999.
8. Alla Zusman, Boris Zlotin, Gafur Zainiev, An Application of Directed Evolution, http://www.Ideation-triz.com, June, 2001.
9. Tan Runhua, A Macro Process Model for Product Innovation Using TRIZ, the eighth world conference on integrated design and process technology, Beijing, China, 2005.

# Axiom-based Potential Functional Failure Analysis for Risk-free Design

Lian Benning, Cao Guozhong, Bai Zhonghang, and Tan Runhua
Hebei University of Technology, Department of Mechanical Engineering,
Guang Rong Road NO.8, PO Box: 165, 300130 Tianjin, P.R. China
Phone: +86-22-60204871, Fax: +86-22-26564037
E-mai: lianbenning@126.com

**Abstract.** Quality and reliability of product is not established completely in detail design process, but is brought out essentially in the course of conceptual design. A potential functional failure analysis method to improve reliability of product based on design axiom in the stage of conceptual design was introduced in this paper. This method provided designers with an analytical and non-probabilistic tool to evaluate the result of conceptual design from the opinion of design axioms. Functional failure modes can be identified by function analysis based on the ideality axiom, the independence axiom and the information axiom. These potential functional failures point out working direction for the latter improving design. A speedy cutting off valve in the TRT (Top Gas Pressure Recovery Turbine) system is studied as an example to illustrate this method's potential.

## 1 Introduction

In product design, the end goal is to develop a product which performs a function or functions to satisfy the customer's needs. Pahl and Beitz [1] state that nearly 80% of the costs and problems are created in product development and that cost and quality are essentially designed into products in the conceptual stage. As to produce a more reliable product without the need for multiple redesigns, it is so significant to begin the failure analysis of a product design to prevent failure modes in advanced stage of the design process. Classical failure analysis techniques such as Failure Modes and Effects Analysis (FMEA), Anticipatory Failure Determination (AFD) and Fault Tree Analysis (FTA) are used in currently industries to determine potential failures of products [2-4]. In order to eliminate or reduce the possibility of failure, designers need to be aware of all of the potential significant failure modes in the systems being designed. An essential and crucial part of these methods is a required function-failure knowledge base of previous products. Tumer and Stone [5] collected historical failure data to establish a connection between failure modes and product's

*Please use the following format when citing this chapter:*

Benning, L., Gouzhong, C., Zhonghang, B., Runhua, T., 2007, in IFIP International Federation for Information Processing, Volume 250, Trends in Computer Aided Innovation, ed. León-Rovira, N., (Boston: Springer), pp. 107-114.

functions. Also some other researchers did a considerable amount of work in developing the function-failure knowledge base in their respective field [6-8]. In this paper, an axiom-based functional failure analysis method for risk-free design is presented in functional design of product to change the experience-based functional failure analysis into the use of scientific theories and methodologies that based on design axioms.

There are two ways to deal with design: experiment-based and axiom-based. In essence, classical failure analysis techniques mentioned above are experiment-based approaches because they needed a mass of failure data which should be collected in real application or experiments. The experiment-based approach is generally useful at the detail level. However, it is difficult to match failure data to product functions very well, for there is no detail structure information of product in the conceptual design, especially in functional design. The axiom-based approach to design is based on the abstraction of good design decisions which is easy to be reused in the conceptual design process.

## 2 Design Axioms

Axioms are established rules, principles, or laws. Fields such as mathematics, physical science, and engineering have gone through the transition from experience-based practice to the use of scientific theories and methodologies that are based on axioms. Altshuller has introduced the ideality axiom in the theory of inventive problem solving (TRIZ) [9]. Suh has presented two design axioms in his axiomatic design approach [10]. The first design axiom is known as the independence axiom and the second axiom is known as the information axiom. All these design axioms are stated as follows:

- Axiom 1: The Ideality Axiom

Since technological system are designed to perform certain function or functions, a better system obviously requires less material to built and maintain, and less energy to operate, to perform these functions. Altshuller pointed out that an ideal technological system is a system whose mass, dimensions, cost, energy consumption, etc. are approaching zero, but whose capability to perform the specified function is not diminishing [9]. The ideality axiom states that the design with the highest ideality is the best design. The ideality of a product or a technical system is one of the major criterions when we assess the quality of design.

- Axiom 2: The Independence Axiom

In axiomatic design, the design objective is defined in terms of functional requirements (FRs) and the physical solution is characterized in terms of design parameters (DPs). Design task can be represented as a mapping between functional domain and physical domain as show in Fig. 1. The independence axiom states that mapping between the FRs and DPs in an acceptable design is such that each FR can be satisfied without affecting any other FRs.

- Axiom 3: The Information Axiom

The information axiom states that the design with the highest probability of achieving the design goals as expressed by the functional requirements (FRs) is the best design. The system performance is shown by its probability density function (pdf), which defines the system range. The FR is satisfied only when the system rang is within the design range. The overlap between design rang and system range is called common range, which is the only region where the FR is satisfied as shown in Fig. 2.

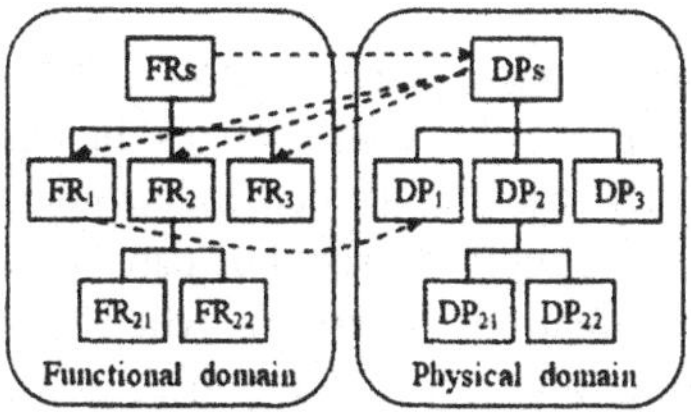


**Fig. 1.** Zigzagging to decompose FRs and DPs in the functional and physical domains to create the FRs and DPs hierarchies.

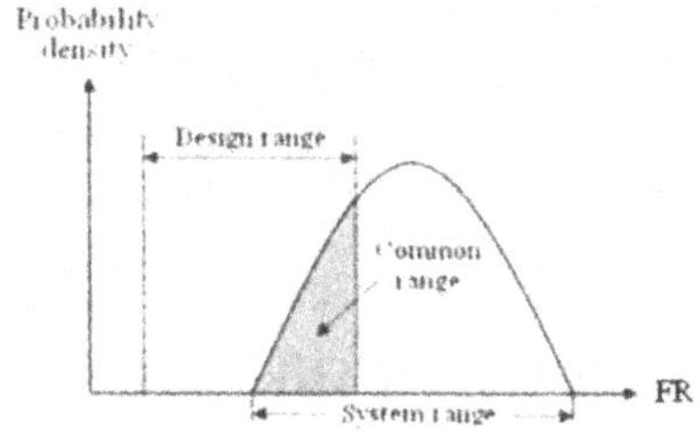


**Fig. 2.** Design range, system range, common range and pdf for an FR.

# 3 Axiom-based Functional Failure

## 3.1 Functional Failure based on Axiom 1

One principle of TRIZ, and a method of determining the progress is the ideality axiom [11]. Ideality is summarized by the equation:

$$\text{Ideality} = \sum UF / \sum HF \tag{1}$$

where $\sum$ UF is the sum of all useful functions and $\sum$ HF is the sum of all harmful functions. In fact, a technique is a "fee" for realization of the required function, because a technique always has harmful functions. For example, when we have our car braked by brakes, the noises and heat what we do not need will be produced. If the degree of the ideality is too low, that will cause the whole system to fail.

### 3.2 Functional Failure based on Axiom 2

In axiomatic design, the mapping between functional domain and physical domain can be described mathematically by the following equation:

$$\{FR\} = |A|\{DP\} \quad (2)$$

where {FR} is the functional requirement vector, {DP} is the design parameter vector, and |A| is the design matrix that characterizes the design. When the |A| matrix is a diagonal matrix, each DP can be decided independently by its respective FR. When the |A| matrix is a triangular matrix, DP can be decided independently by following the order of determination. Otherwise, when the |A| matrix has no special structure, one function failing to satisfy its DP will cause many other functions to be performed unsuccessfully.

### 3.3 Functional Failure based on Axiom 3

The information axiom indicates that the variance of the system must be small and the bias must be eliminated to make the system range lie inside the design range for enhancing robust of the design. The insufficient useful action or the excessive useful action of the FR will cause the respective function to fail when the bias between the system range and the design range overruns the limit.

## 4 Case Study: a Speedy Cutting Off Valve

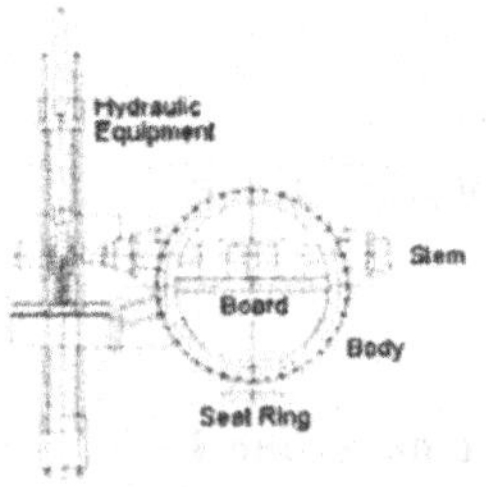


**Fig. 3.** Structure of the speedy cutting off valve.

In order to use excessive blast furnace gas, a kind of equipment called TRT (Top Gas Pressure Recovery Turbine) is developed. Speedy cutting off valve is used for urgent cutting off the TRT, as shown in Fig. 3. From the point of view of system function, the Fig. 4 shows the problems of the speedy cutting off valve in TRT. Three kinds of functional failure modes can be identified based on design axioms. The harmful actions: the dust polluting the oil, the filter and the pipeline counteracting the fluid of oil, the hydraulic cylinder counteracting the movement of the piston; The insufficient action: the dynamical spring driving the piston deficiently; The excessive actions: the motor and the hydraulic control box consuming massive electricity supply.

All these functional failures mentioned above point out working direction for the latter improving design. The improved function model of the speedy cutting off valve is shown in Fig. 5. In the improved drive system the hydraulic devices are removed completely and a new drive device is adopted shown in Fig. 6. The system can realize all motions, such as the slow opening, slow shutting, quick shutting and moving about movements, as well as the buffering and the complete closure of valve.

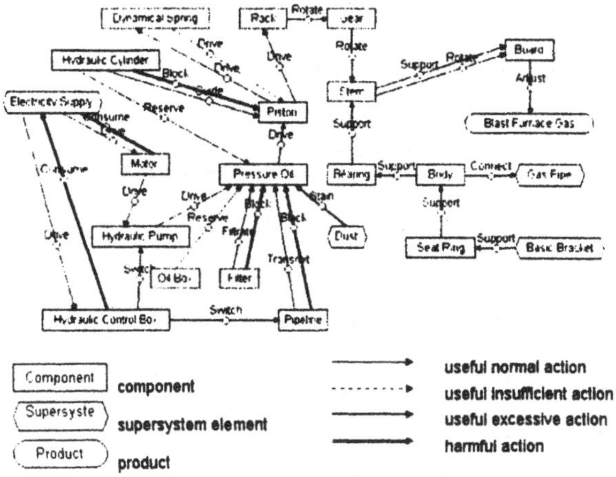


**Fig. 4.** Structure of the speedy cutting off valve.

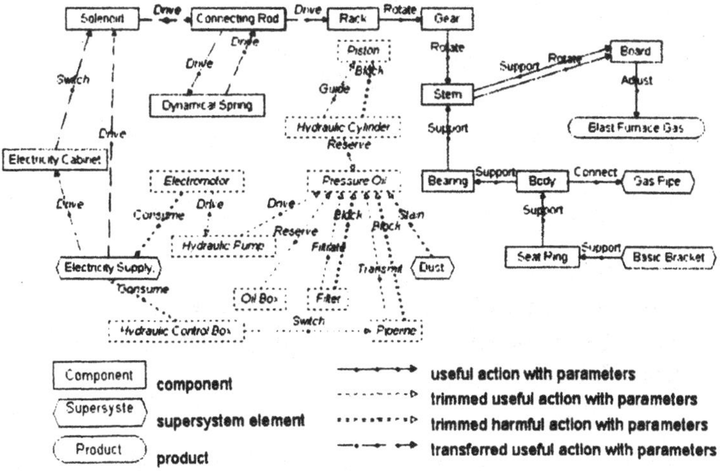


**Fig. 5.** Improved Function Model of the Speedy Cutting Off Valve

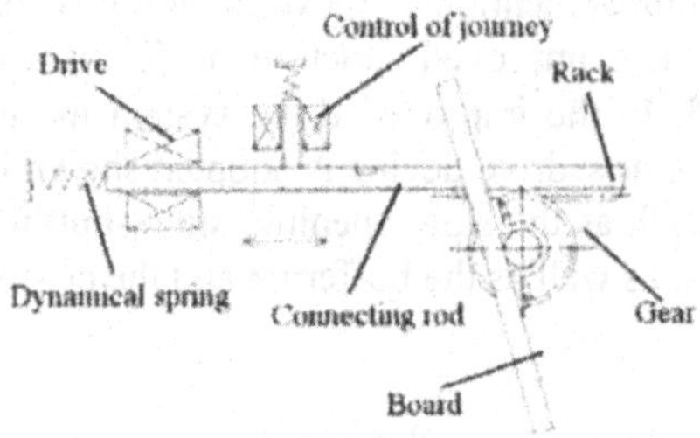


**Fig. 6.** Principle Solution of the New Drive System

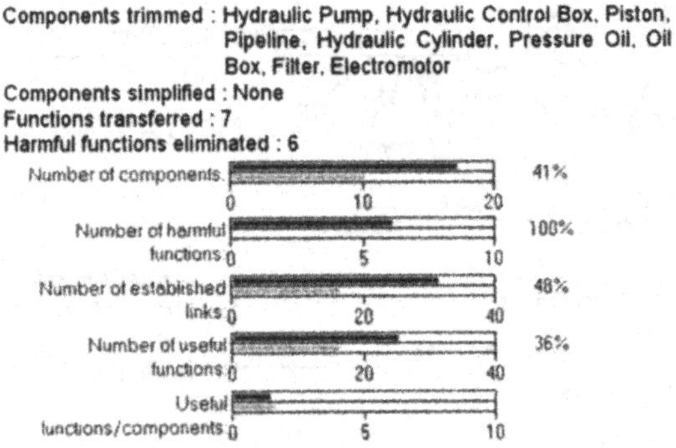


**Fig. 7.** Improvement Result of the Speedy Cutting Off Valve

The FRs and DPs of the new drive system as shown in Fig. 6 are:

$FR_1$: supply power $DP_1$: power device

$FR_2$: transfer power $DP_2$: transfer device

$FR_3$: control power $DP_3$: control device

The relation of FRs to DPs is:

$$\begin{Bmatrix} FR_1 \\ FR_2 \\ FR_3 \end{Bmatrix} = \begin{bmatrix} 1 & 0 & 0 \\ 1 & 1 & 0 \\ 1 & 1 & 1 \end{bmatrix} \begin{Bmatrix} DP_1 \\ DP_2 \\ DP_3 \end{Bmatrix} \tag{3}$$

The $FR_1$ and $DP_1$ can be more divided into:

$FR_{11}$: supply closing power $DP_{11}$: closing device

$FR_{12}$: supply opening power $DP_{12}$: opening device

The relation of FRs to DPs is:

$$\begin{Bmatrix} FR_{11} \\ FR_{12} \end{Bmatrix} = \begin{bmatrix} 1 & 0 \\ 1 & 1 \end{bmatrix} \begin{Bmatrix} DP_{11} \\ DP_{12} \end{Bmatrix} \tag{4}$$

According to the two design equations, the requirements satisfy the Independence Axiom. There is no coupled functional failure in the new driving system of the improved speedy cutting off valve.

In this case, comparing the improved system with the original system as shown in Fig. 7, 33% of elements and 43% of interactions between elements are decreased, potential functional failures are eliminated completely, the system structure is simplified and the cost is reduced. From the view of function, the negative effects aroused by the hydraulic system, such as the pollution of oil, the flow resistance of oil, the unreliability of hydraulic meter, deficient force of spring and etc., are removed in the original system by substituting the hydraulic drive system with the new electromagnetic drive system. Every movement of the speedy cutting off valve can be implemented well, the operation is much simpler and the system performance is improved greatly.

## 5 Conclusion

Using the idea of axiom, in a larger way than the idea of statistics, it is possible to apply the functional failure analysis in the conception design when there is lack of function-failure data collected by a great lot of failure reports or reliability experiments. In this paper, a functional failure analysis method based on design axioms was introduced to improve reliability and performance of products in the stage of conceptual design.

According to the ideality axiom, all the harmful function should be removed to increase the ideality of the system. The independence axiom indicates that the coupled design is undesirable, for the change in any DP may cause many FRs to fail simultaneously. The information axiom points out that the insufficient useful action or the excessive useful action of the FR will cause the respective function to fail when the bias between the system range and the design range overruns the limit. Therefore, we identify and then eliminate all these potential functional failures when functional model is constructed to improve the reliability and performance of products.

In the case study of the speedy cutting off valve in the TRT, comparing the improved system with the original system proves that this method is feasible.

## 6 Acknowledgment

This research is supported in part by the Natural Science Foundation of China under Grant Numbers 50675059 and the National High Technology Research and Development Program of China under Grant Numbers 2006AA042109. Any opinions or findings of this work are the responsibility of the authors, and do not necessarily reflect the views of the sponsors or collaborators.

## 7 References

1. G. Pahl and W. Beitz, *Engineering Design: A Systematic Approach* (Springer Verlag, 1996).
2. A. Carter, *Mechanical Reliability and Design* (Wiley, 1997).
3. E. Henley and H. Kumamoto, Probabilistic Risk Assessment: Reliability Engineering, *Design and Analysis,* IEEE (1992).

4. I. Tumer and R. Stone, Analytical Method for Mapping Function to Failure during High-risk Component Development, in: *ASME Design for Manufacturing Conference*, DETC2001/DFM-21173, Pittsburgh, PA (2001).
5. I.Y. Tumer and R.B. Stone, Mapping Function to Failure during High-risk Component Development, *Research in Engineering Design*, 14(1), 25-33 (2003).
6. S.G. Arunajadai, R.B. Stone, and I.Y. Tumer, A Framework for Creating a Function-based Design Tool for Failure Mode Identification, in: *Proceedings of the 2002 ASME Design Engineering Technical Conference, Design Theory and Methodology Conference*, Montreal, Canada (2002).
7. R.A. Roberts, R.B. Stone, and I.Y. Tumer, Deriving Function-Failure Information for Failure-free Rotorcraft Component Design, in: *Proceedings of the 2002 ASME Design Engineering Technical Conferences, Design for Manufacturing Conference*, DETC2002/DFM-34166, Montreal, Canada (2002).
8. I.Y. Tumer, R.B. Stone, R.A. Roberts, and A.F. Brown, A Function-based Exploration of JPL's Problem/Failure Reporting Database, *Proceedings of the 2003 ASME International Mechanical Engineering Congress and Expo*, IMECE2003-42769, Washington (2002).
9. V. Fey and E.I. Rivin, *Innovation on Demand: New Product Development Using TRIZ* (Cambridge University, 2005).
10. N.P. Suh, *The Principles of Design* (Oxford University, 1990).
11. S.D. Savransky, *Engineering of Creativity: Introduction to TRIZ Methodology of Inventive Problem Solving* (CRC, 2000).

# Using the CAI software (InventionTool 3.0) to solve complexity problem

Zhang Peng, Tan Runhua
School of Mechanical Engineering, Hebei University of technology ,
Tianjin, 300130, People's Republic of china,
landi1979@sina.com, rhtan@hebut.edu.cn

**Abstract.** InventionTool 3.0, which is a CAI software based on TRIZ, is created by Institute of Innovation of Hebei University of' Technology. Using the CAI software, designers can solve innovation problem. During the design process, the InventionTools 3.0 plays a very important role. Firstly, the technical evolution module of the InventionTools 3.0, as an effective method, is used to obtained conceptual design and several solutions. Secondly, using the Complexity Theory based on the Axiomatic Design, the solutions are analyzed. Finally, according to analysis result, the best solution would be obtained. In this paper, the pill counting system is used as an complexity problem to show how the InventionTool 3.0 and the Complexity Theory based on the Axiomatic Design is applied.

## 1 Introduction

The dropping pill is a Chinese traditional medicine and has won the favor of people all over the world. For the dropping pill is quite unique, some existing production line is unable to satisfy the requirements of its production. A new production line is needed, which includes three sections: the dropping pill machine, the scratching pill machine and the packaging machine. For reducing the cost and improving the competition, we pay attention to increasing the quality of design. Using the InventionTools 3.0 is an important way to achieve it. During the design process, the InventionTools 3.0 plays a very important role. Firstly, the technical evolution module of the InventionTools 3.0, as an effective method, is used to obtained conceptual design and several solutions. Secondly, using the Complexity Theory based on the Axiomatic Design [1], the solutions are analyzed. Finally, according to analysis result, the best solution would be obtained.

The pill counting system, which has shown the application of the law and routes of system evolution, is a sub-system in the production line of dropping pill. During

---

*Please use the following format when citing this chapter:*

Peng, Z., Runhua, T., 2007, in IFIP International Federation for Information Processing, Volume 250, Trends in Computer Aided Innovation, ed. León-Rovira, N., (Boston: Springer), pp. 115-123.

the design process, the InventionTools 3.0 is used. What is the InventionTools 3.0? How did we use the InventionTools 3.0? Why do we apply the Complexity Theory based on the Axiomatic Design?

## 2 Preliminaries

### 2.1 The TRIZ technology forecasting

The TRIZ technology forecasting [2] developed by Altshuller [3] is a set of patterns and paths, which show the trends of technological systems evolution in structures. These patterns and paths are revealed by analysis of hundreds of thousands of invention descriptions available in the world patent databases. The most important finding is that the patterns and paths revealed in one engineering field can be transferred to other kinds of artificial systems [4]. The law and routes of system evolution are described that the routes of system evolution are the order of structural evolution of the core technology. All in all, we can obtain the trends of technological systems by the TRIZ technology forecasting.

When we develop a new product, it is necessary for the function units to be described precisely and detailedly and for the system model to be established. The chief task for designers is to confirm the function units, analyze the correlation among the function units and establish the function model of system. In order to realize the demand mentioned above, function analysis is adopted. According to technology evolution, several solutions are obtained when Invention Tool 3.0 is used. If the function unit is independent, the best could be obtained in the several solutions. On the other hand, if the function unit is not independent, the best might be hidden in the several solutions. A system method is needed to obtain design optimization. The Complexity Theory based on the Axiomatic Design provides the theoretical basis for design optimization.

### 2.2 Complexity Theory based on the Axiomatic Design

In 1999 N.P. Suh put forth the Complexity Theory based on the Axiomatic Design method. What is complexity? How do we reduce complexity? Why do people think that a product with many parts is complex? Complexity is measured in the functional domain rather than in the physical domain, which distinguishes this theory from other complexity theories and yields a unique and inherent perspective on complexity [5].

The design effort may produce several designs, all of which may be acceptable in terms of the technology evolution. It is likely that different designers will come up with different designs, because there can be many designs that satisfy the function requirement [5]. However, one of these designs is likely to be superior to the others. The Complexity Theory based on the Axiomatic Design is useful in selecting the best among those designs that are acceptable. Among the designs that acceptable from the functional point of view, one may be superior to others in terms of achieving the design goals as expressed by the functional requirements. The Complexity Theory based on the Axiomatic Design states that the design with the highest probability of success is the best design.

Complexity is defined as a measure of uncertainty in achieving the functional requirements (FRS) of a system [6]. A design is called complex when its probability of success is low, that is, when the information content required to satisfy the FRs is high. A physically large system is not necessarily complex if the information content is low. Conversely, even a small system can be complex if the information content is high [6].

## 2.3 InventionTool 3.0

InventionTool 3.0, which is a CAI software based on TRIZ, is created by Institute of Innovation of Hebei University of Technology. Figure 1 shows the homepage of the InventionTool 3.0.

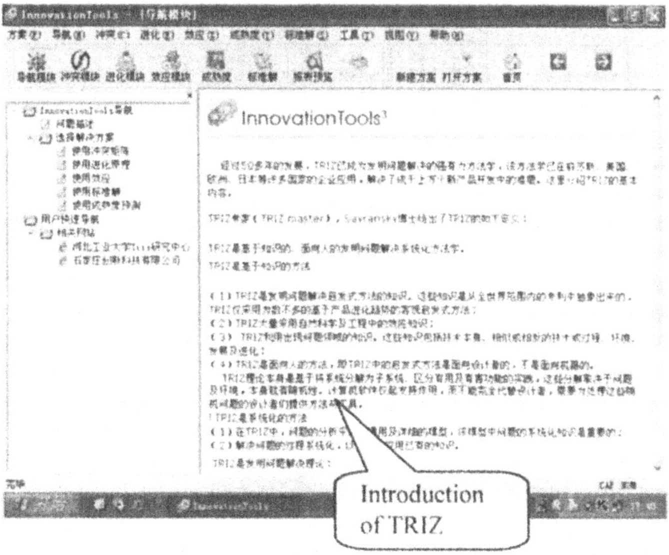


**Fig.1.** The homepage of the Invention Tool 3.0

At present, CAD software can be used in the technical design and the detailed design, but can not be used in the conceptual design. The CAD software can not be used in helping the designer to obtain innovation plans. Using the CAI software, designers can solve the “innovation problem”.

There are three modules in the InventionTool 3.0: the evolution module, the conflict module and the effect module. According to these modules, designers can obtain several solutions.

The InventionTool 3.0 solves "innovation question" based on TRIZ [3]. TRIZ was developed in Russia by Genrich Altshuller, a talented scientist and inventor, and his followers. Altshuller's work with TRIZ began in the 1940s and, to date, much experience in applying TRIZ application to various areas of human activity has been amassed [3]. TRIZ is based on the study and application of the patterns of evolution of various systems-technological machines, manufacturing processes, scientific theories, organizations, works of art, and so on. Based on these patterns, methods have been developed for searching for creative solutions. TRIZ includes several parts: TRIZ tools, repository and technology forecasting [4]. A person masters all of them is difficult. So, the computer aided innovation software is necessary. For cutting the designing time and improving the level of innovation, designers pay attention to using the computer aided innovation software. Figure 2 shows the homepage of the evolution module.

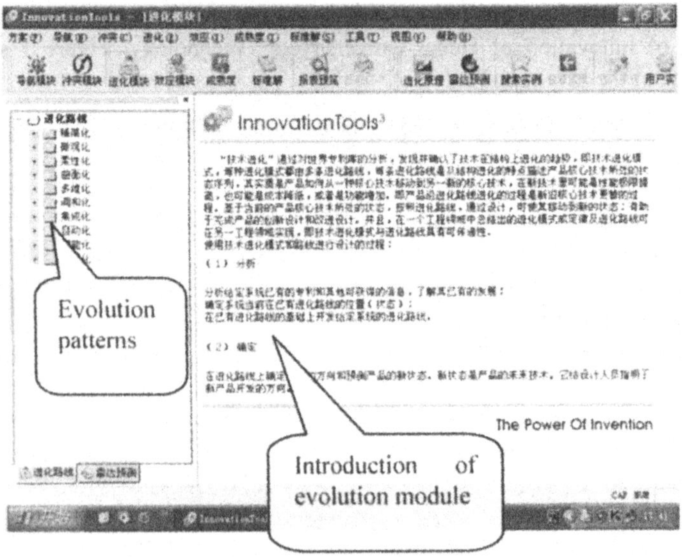


**Fig.2.** The homepage of evolution module.

# 3 Case study

The design process of the pill counting system is a typical example of the whole design process. In the process, InventionTool 3.0 is used. The pill counting system is divided into two parts: the pill boards and counting sensors.

The principle of the pill counting system is: the sensor send out rays which can be reflected by pills. If the sensor doesn't receive the ray which is sent out by itself and reflected by the pill, indicate there is not a bill. Figure 3 shows the principle of the pill counting system.

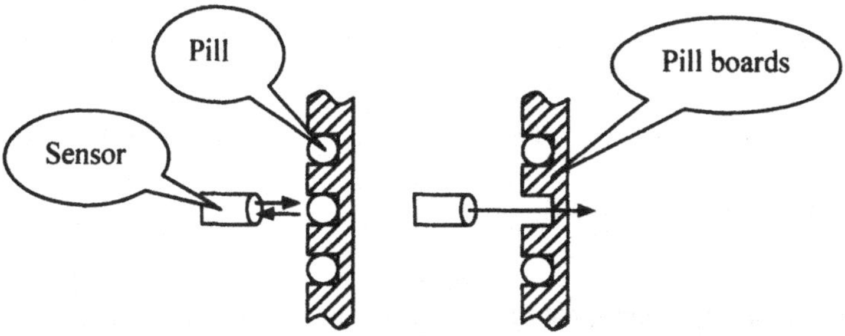


**Fig.3.** The principle of the pill counting system.

### 3.1 Design the arrangement of the counting sensors

According to the evolution path “Mono – bi – poly” [8] of the InventionTool 3.0, several solutions were obtained when the function requirement of the pill counting system was known. Figure 4 shows the evolution path “Mono – bi – poly” of the InventionTool 3.0.

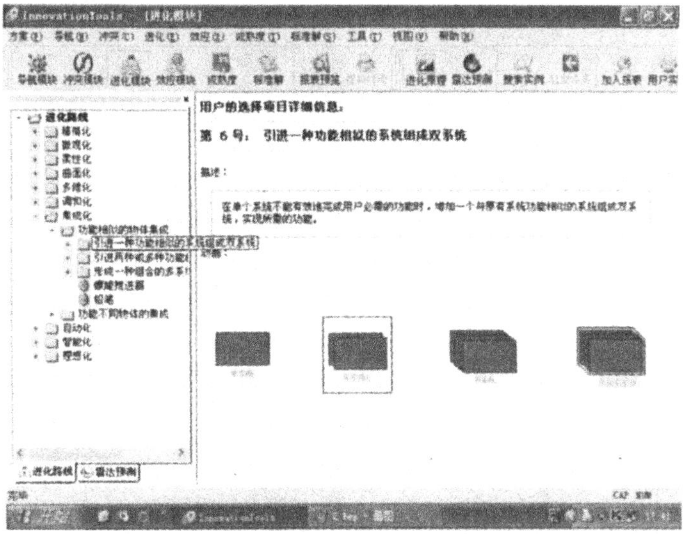

**Fig.4.** The evolution path “Mono – bi – poly”

Using the evolution path “Mono – bi – poly” of the InventionTool 3.0, the arrangement of the counting sensors can be designed as figure 5.

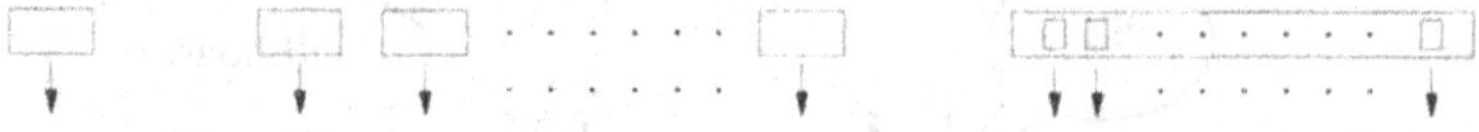

(a) (b) (c)

**Fig.5.** Three projects of the counting sensors.

Now, let's analyze the three projects using the complexity theory based on the Axiomatic Design method. When the project one is adopted, we will see a single system with only one sensor. In this project, the counting system's function can be achieved easily. But the project is inefficient, so it is not available. According to the evolution path" Mono – bi – poly", we can choose multi-system. Fig.6-b shows the project two. Several same sensors are used. The project two is more efficient than the project one. But the sensors' operational complexity may be increased. This project is not available, too. Is there a project which will increase efficiency but not complexity? Project three, which is shown in fig.6-c, can achieve it. The sensors are integrated. Efficiency is improved, and operational complexity is not increased.

We apply the three projects in counting system and analyze their feasibilities. We regard the cylinder as the pill board.

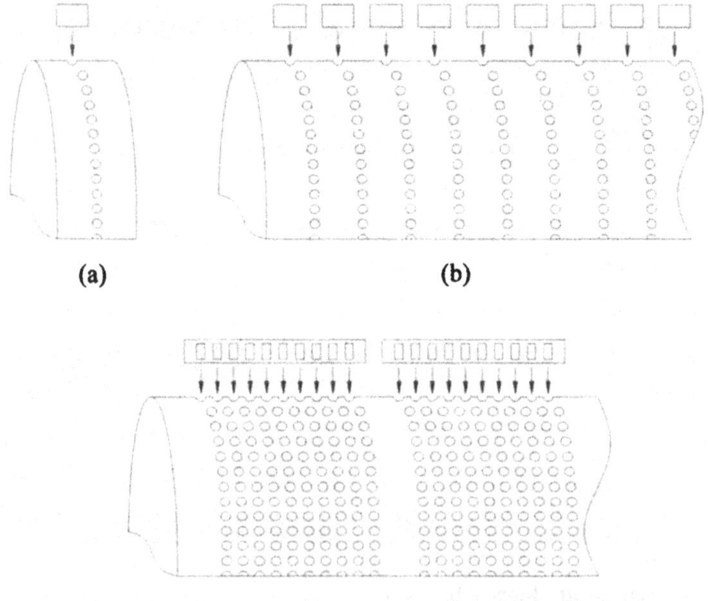

(a) (b)

(c)

**Fig.6.** Three projects of the counting sensors

Fig.6-a shows the project one, there is only one set of holes are used to put pills. It is easy to design and manufacture this system. But this system is not available for its inefficient.

Fig.6-b shows the project two, this system involves several sensors and several sets of holes. Although efficiency is improved, operational complexity is increased simultaneously. Besides, complexity of installation is also increased.

Fig.6-c shows the project three. For reducing the complexity and improving the efficiency, the sensors are integrated. Is the project three better than project two? By integrated, the size of the sensors will be reduced, and the space between two sets of pill holes will be reduced, too. Thus, the pill board's manufactural complexity may be increased.

### 3.2 Design the pill boards

To sum up, we need to complete the design of the pill boards to gain a satisfying counting system. We design the pill boards based on the evolution module of the InventionTool 3.0 correspond with the design of the sensor.

When we design the pill board, the following question must be considered. How to improve efficiency? Increasing the size of the cylinder is an important way to achieve it, but it will increase the complexity of the system. The evolution path "segmentation of objects and substances" [9] of the InventionTool 3.0, as an effective method, is used to obtain the solutions of the problems.

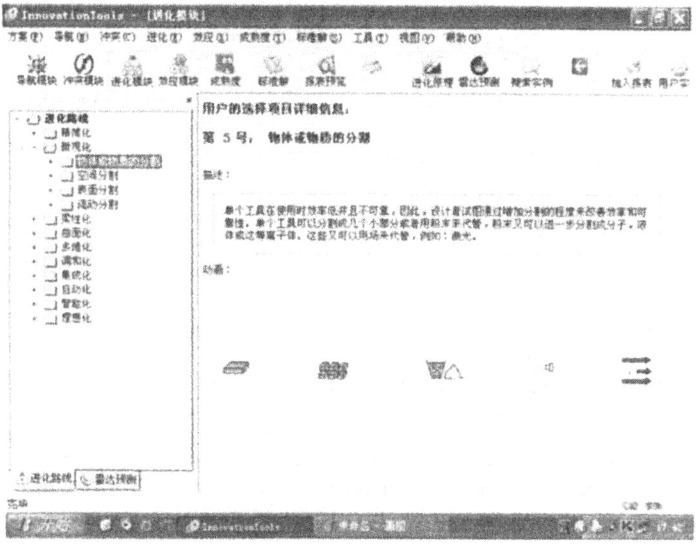

**Fig.7.** The evolution path "segmentation of objects and substances"

According to the evolution path "segmentation of objects and substances", the design has smaller information content when the cylinder is divided into several parts. That is to say, the system's complexity is reduced. The system in the project one only has one sensor and a set of pill hole, so its cylinder can't be divided into several parts. We analyze the project two and project three.

The cylinder in the project two is divided into several parts. It is shown in fig. 8.

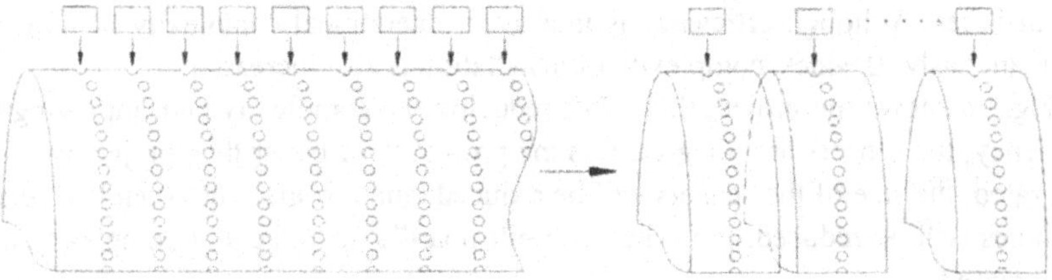

**Fig.8.** Improving the project two

The manufactural complexity is reduced, but many other complexities, such as installation complexity come forth with it. The complexity of the counting system is not improved evidently. The project two is not available.

As is shown in fig.9, the cylinder in the project three is divided into several parts. The manufactural complexity is reduced, and it's convenient to install. The project three is efficient with smaller complexity. The project three is the best design.

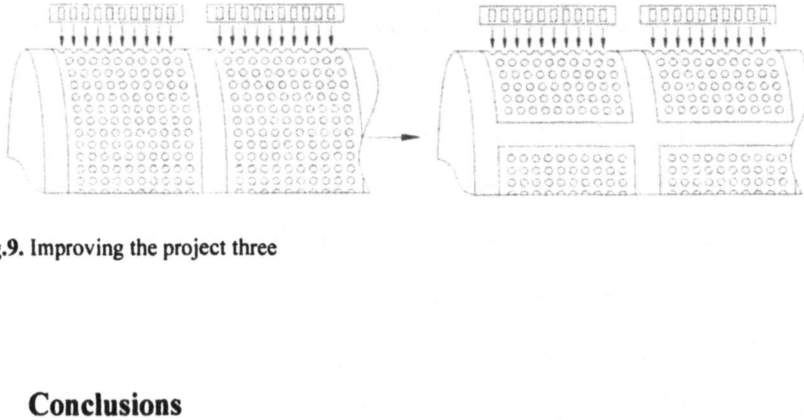

**Fig.9.** Improving the project three

# 4 Conclusions

The InventionTool 3.0 can be used to support designers to find several projects. The best is usually hidden and designers do not know it. According to the Complexity Theory based on the Axiomatic Design method, designers can find it and implement it in the design of products. Firstly, some projects are raised by the design of the pill counting sensor. Secondly, the projects are analyzed for their feasibilities based on the Complexity Theory. Lastly, the smallest information content project is obtained, so it is the best.

The CAI software plays a very important role in design. Designers pay attention to the computer aided innovation software. As the first domestic CAI software in china, InventionTool 3.0 has won the favor of national enginery company, and has been already used by several companies. As is shown in the process of designing the pill counting system, the Invention Tool 3.0 may provide many design projects for the designer, but the choice of the projects still depend on the designer oneself. The

InventionTool 3.0 can't do it .The tendency of the Invention Tool x software is adding the optimization module based on the Complexity Theory.

## 5 Acknowledgements

The research is supported in part by the Chinese Natural Science Foundation under Grant Numbers 50675059 and Chinese national 863 planning project under Grant Number 2006AA04Z109. No part of this paper represents the views and opinions of any of the sponsors mentioned above.

## 6 References


1. 1 Suh N.P. Axiomatic Design: Advances and Applications (Oxford University Press, New York, 2001).
2 Suh N.P.The Principles of Design(Oxford University Press, New York, 1990).
3 Altshuller G, The Innovation Algorithm, TRIZ, systematic innovation and technical creativity, Technical Innovation Center, INC., Worcester, 1999
4 Savransky S D, Engineering of creativity, CRC Press, New York, 2000
5 Suh N P. "A theory of complexity, periodicity, and design axioms", Research in Engineering Design, Vol.11, 1999, pp. 116-131.
6 Suh N P. Design-in of Quality through Axiomatic Design. IEEE Transactions on Reliability, 44(2): 1995.
7 TAN Runhua. A macro process model for product innovation using TRIZ. The eighth world conference on integrated design & process technology, Beijing, China, 2005, 766-773
8 Twiss B., Managing technological innovation, Fourth edition, Pitman Publishing, London, 1992
9 Twiss B., Forecasting for technical decisions, Peter Peregrinus, London, 1992
10 Fey V.R. and Rivin E.I., Guided technology evolution (TRIZ Technology Forecasting), http://www.trizgroup.com/article2.htm l
11 Zusman A., Zlotin B., Zainiev G., An application of directed evolution, http://www.ideationtriz.com/Endoscopic_case_study.htm

# Generative Art Images by Complex Functions Based Genetic Algorithm

Hong Liu, Xiyu Liu
School of Information Scinece and Engineering, Shandong Normal University, Jinan, P. R. China,250014
School of Management, Shandong Normal University, Jinan, , P. R. China,250014
lhsdcn@jn-public.sd.cninfo.net

**Abstract.** This paper presents a novel computer supported design system which uses computational approach to producing 3D images for stimulating creativity of designers. It put forward generic algorithm based on a binary tree to generate 3D images. This approach is illustrated by an artwork design example, which uses general complex function expressions to form 3D images of artistic flowers. It shows that approach is able to generate some innovative solutions and demonstrates the power of computational approach.

## 1 Introduction

The design research community has spent much of its effort in recent years developing computer supported design systems. Generative design systems - systems for specifying, generating and exploring spaces of designs and design alternatives - have been proposed and studied as a topic of design research for many years.

Generative Design is an excellent snapshot of the innovative process from conceptual framework through to specific production techniques and methods. It is ideal for aspiring designers and artists working in the field of computational media. Especially those who are interested in the potential of generative/ algorithmic/ combinational/ emergent/ visual methods and the exploration of active images.

This paper presents a novel computer supported design system that uses evolutionary approach to generate 3D images. The tree structure based genetic algorithms and complex functions are used in this system. Programs are implemented by using Visual C++6.0 and mathematical software MATLAB.

The remainder of this paper is organized as follows. Section 2 is related work on generative design. Section 3 introduces the tree structure based genetic algorithm. In section 4, an artwork design example is presented for showing how to use the tree structure based genetic algorithm and complex functions to generate 3D images.

*Please use the following format when citing this chapter:*

Liu, H., Liu, X., 2007, in IFIP International Federation for Information Processing, Volume 250, Trends in Computer Aided Innovation, ed. León-Rovira, N., (Boston: Springer), pp. 125-134.

Finally, these results are briefly analyzed, followed by a discussion of future improvements.

## 2 Related work

Generative design is the idea realized as genetic code of artificial objects. The generative project is a concept-software that works producing three-dimensional unique and non-repeatable events as possible and manifold expressions of the generating idea identified by the designer as a subjective proposal in a possible world. This idea/human creative art renders explicit and realizes an unpredictable, amazing and endless expansion of human creativity. Computers are simply the tools for its storage in memory and execution [1].

Generative design describes a broad class of design where the design instances are created automatically from a high-level specification. Most often, the underlying mechanisms for generating the design instances in some way model biological processes: evolutionary genetics, cellular growth, etc. These artificial simulations of life processes provide a good conceptual basis for designing products.

Evolving design concepts by mutating computer models in a simulated environment is now a well-established technique in fields as diverse as aeronautics, yacht design, architecture, textile design, fine art and music [2]. Some of the work was performed by Professor John Frazer, who spent many years developing evolutionary architecture systems with his students. He showed how evolution could generate many surprising and inspirational architectural forms, and how novel and useful structures could be evolved [3]. In Australia, the work of Professor John Gero and his colleagues also investigated the use of evolution to generate new architectural forms. This work concentrates on the use of evolution of new floor plans for buildings, showing over many years of research how explorative evolution can create novel floor plans that satisfy many fuzzy constraints and objectives [4]. They even show how evolution can learn to create buildings in the style of well-known architects.

In Argenia, a system for architectural design by Soddu, the three-dimensional models produced can be directly utilized by industrial manufacturing equipment like numerically controlled machines and robots, which already represent the present technologies of industrial production. This generative and automatic reprogramming device of robots makes it possible to produce unique objects with the same equipment and with costs comparable to those of objects that are identical; like a printer that can produce pages that are all the same or all different, at precisely the same cost [5].

The list of experimental application of generative techniques to physical design also includes airfoil optimization [6], clothing design [7], and an experimental test-bed for consumer controlled generative product design by Emergent Design [8].

However, the development of generative design tools is still at its early stage. The research and development of design support tools using evolutionary computing technology are still in process and have huge potential for the development of new design technology.

## 3 The tree structured genetic algorithm

General generic algorithms use binary strings to express the problem. It has solved many problems successfully. But it would be inappropriate to express flexible problem. For example, mathematical expressions may be of arbitrary size and take a variety of forms. Thus, it would not be logical to code them as fixed length binary strings. Otherwise the domain of search would be restricted and the resulting algorithm would be restricted and only be applicable to a specific problem rather than a general case. Thus, tree structure, a method useful for representing mathematical expressions and other flexible problems, is presented in this paper.

For a thorough discussion about trees and their properties, see [9,10]. Here, we only make the definitions involved in our algorithm and these definitions are consistent with the basic definitions and operations of the general tree.

**Definition 1** A binary complex function expression tree is a finite set of nodes that either is empty or consists of a root and two disjoint binary trees called the left sub-tree and the right sub-tree. Each node of the tree is either a terminal node (operand) or a primitive functional node (operator). Operand can be either a variable or a constant. Operator set includes the standard operators (+, -, *, /, ^ ) , basic mathematic functions (such as sqrt (), exp( ), log( ) ), triangle functions ( such as sin(x), cos(x), tan(x),cot(x),sec(x),csc(x),asin(x),acos(x)), hyperbolic functions (such as sinh( ), cosh( ), tanh ( ), asinh ( ), acosh( ), atanh( ) ), complex functions (such as real(z), imag(z), abs(z), angle(z), conj(z)) and so on.

Here we use the expression of mathematical functions in MATLAB (mathematical tool software used in our system).

A binary complex function expression tree satisfies the definition of a general tree:

There is a special node called the root.

The remaining nodes are partitioned into $n \geq 0$ disjoint sets, where each of these sets is a tree. They are called the sub-tree of the root.

Genetic operations include crossover, mutation and selection. According to the above definition, the operations are described here. All of these operations take the tree as their operating object.

(1) Crossover

The primary reproductive operation is the crossover operation. The purpose of this is to create two new trees that contain 'genetic information' about the problem solution inherited from two 'successful' parents. A crossover node is randomly selected in each parent tree. The sub-tree below this node in the first parent tree is then swapped with the sub-tree below the crossover node in the other parent, thus creating two new offspring. A crossover operation is shown as figure 1.

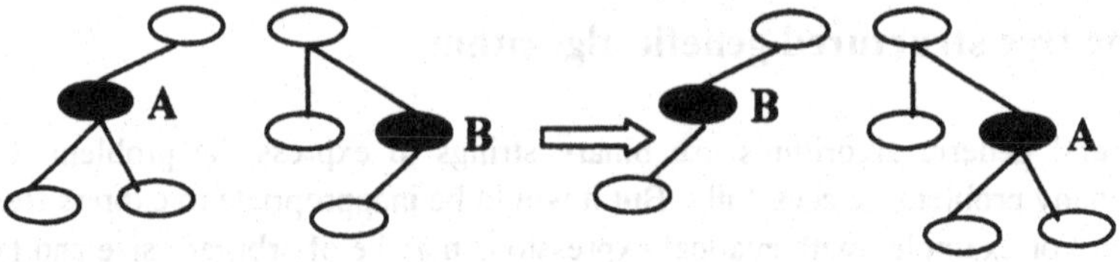


**Fig. 1.** A crossover operation

(2) Mutation

The mutation operation is used to enhance the diversity of trees in the new generation thus opening up new areas of 'solution space'. It works by selecting a random node in a single parent and removing the sub-tree below it. A randomly generated sub-tree then replaces the removed sub-tree. A mutation operation is shown as figure 2.

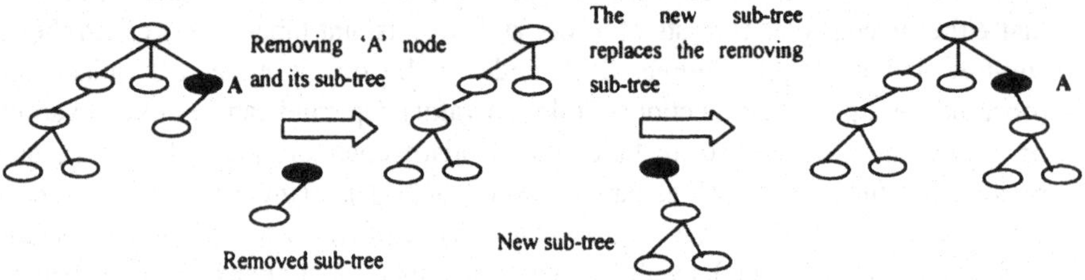


**Fig. 2.** A mutation operation

(3) Selection

For general design, we can get the requirement from designer and transfer it into goal function. Then, the fitness value can be gotten by calculating the similar degree between the goal and individual by a formula. However, for creative design, it has no standards to form a goal function. Therefore, it is hard to calculate the fitness values by a formula. In our system, we use the method of interaction with designer to get fitness values. The range of fitness values is from -1 to 1. After an evolutionary procedure, the fitness values that appointed by designer are recorded in the knowledge base for reuse. Next time, when the same situation appears, the system will access them from the knowledge base[11].

This method gives the designer the authority to select their favored designs and thus guide system to evolve the promising designs. Artificial selection can be a useful means for dealing with ill-defined selection criteria, particularly user centered concerns.

Many explorative systems use human input to help guide evolution. Artists can completely take over the role of fitness function. Because evolution is guided by human selectors, the evolutionary algorithm does not have to be complex. Evolution is used more as a continuous novelty generator, not as an optimizer. The artist is likely to score designs highly inconsistently as he/she changes his/her mind about desirable features during evolution, so the continuous generation of new forms based on the fittest from the previous generation is essential. Consequently, an important element of the evolutionary algorithms used is non-convergence. If the populations of forms were ever to lose diversity and converge onto a single shape, the artist would be unable to explore any future forms.

For clarity, we will present the performing procedure of the tree structured generic algorithms together with a flowers generative design example in the next section.

## 4 A generative artwork example

An artwork design example is presented in this section for showing how to use tree structure based generic algorithm and complex function expressions to generate 3D images.

The complex function expressions are used to produce 3D artistic images. Here, z=x+iy (x is real part and y is virtual part), complex function expression f(z) is an in-order traversal sequence by traversing complex function expression tree.

Both real, imaginary parts and the module of f(z) can generate 3D images by mathematical software MATLAB. Three images of f(z)=sin(z)*log(-z^2)*conj(z) are shown as figure 3.

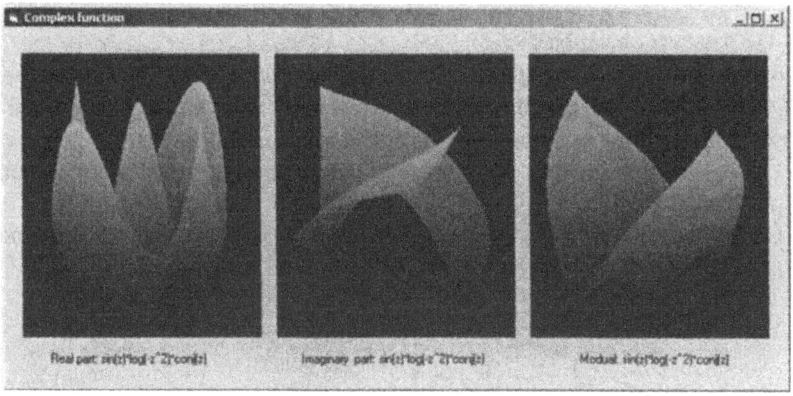


**Fig. 3.** Three images of f(z)=sin(z)*log(-z^2)*conj(z)

Next, we will present the performing process of the algorithm step by step.

Step 1: Initialize the population of chromosomes. The populations are generated by randomly selecting nodes in the set of operands and the set of operators to form complex function expressions. We use the stack to check whether such a complex function expression has properly balanced parentheses. Then, using parsing algorithm, the complex function expressions is read as a string of characters and the binary complex function expressions tree is constructed according to the rules of operator precedence.

Step 2: Get the fitness for each individual in the population via interaction with designer. The populations with high fitness will be shown in 3D form first. Designer can change the fitness value when they have seen the 3D images.

Step 3: Form a new population according to each individual's fitness.

Step 4: Perform crossover and mutation on the population.

Figure 4 shows two complex function expressions trees. Their expressions are f(z)=1.5*sin(z)*cos(-z^2)*angle(-z)*exp(-z) and f(z)=sqrt(z)*log(-z^2)*cot(-z) respectively.

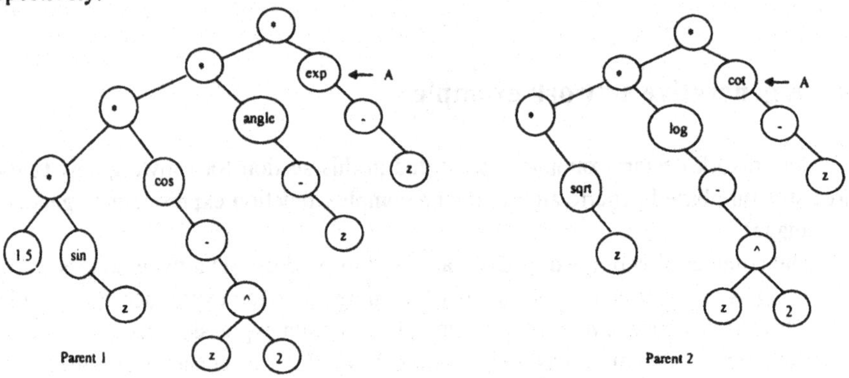


**Fig. 4.** Two parent trees with one crossover nodes

(1) Crossover operation

A crossover node is randomly selected in each parent tree. The sub-tree below this node on the first parent tree is then swapped with the sub-tree below the crossover node on the other parent, thus creating two new offspring. If the new tree can't pass the syntax check or its mathematical expression can't form a normal sketch shape, it will die.

Taking the two trees in figure 4 as parent, after the crossover operations by nodes 'A', we get a pair of children (see figure 5).

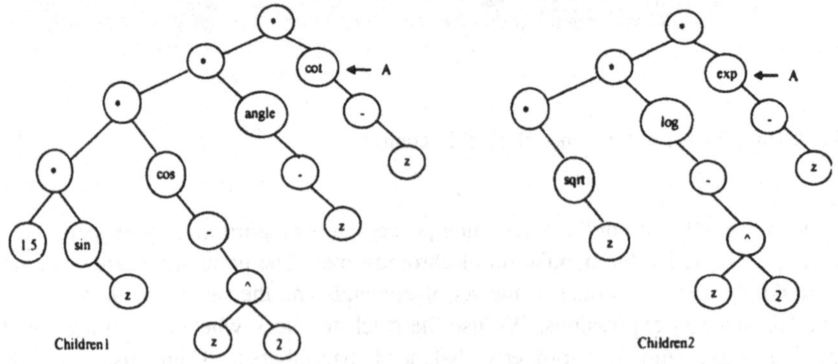


**Fig. 5.** The two children generated by a crossover operation

Figure 6 and figure 7 shows a group of generated 3D images by the module and imaginary part of f(z) correspond to figure 4 and figure 5.

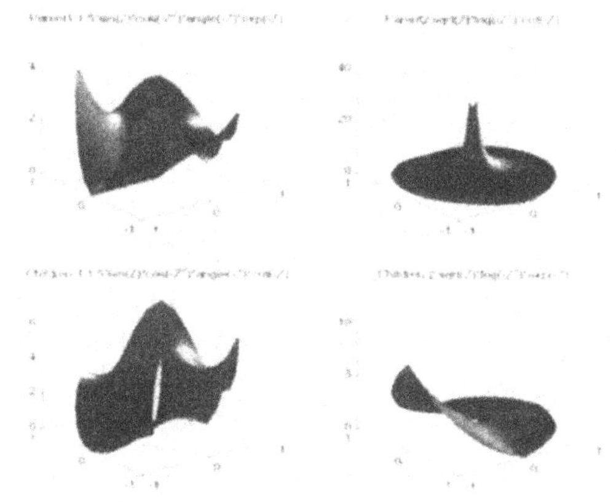

**Fig. 6.** The images correspond to the module of f(z) in figure 4 and figure 5

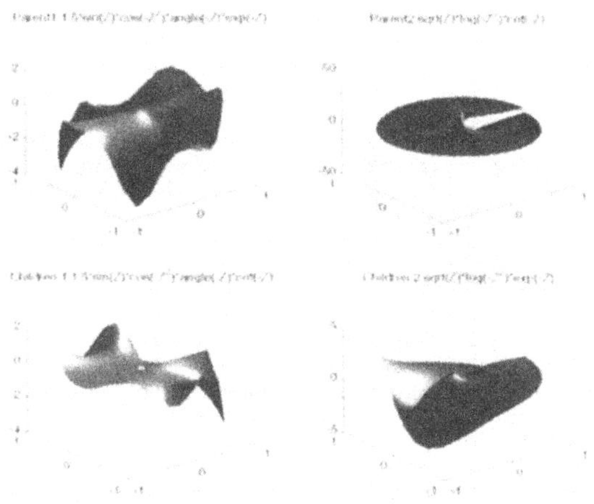

**Fig. 7.** The images correspond to the imaginary part of f(z) in figure 4 and figure 5

(2) Mutation operation

The mutation operation works by selecting a random node in a single parent and removing the sub-tree below it. A randomly generated sub-tree then replaces the removed sub-tree. The offspring will die if it can't pass the syntax check or it can't form a normal shape.

Taking the parent1 tree in figure 4 as a parent, one offspring generated by mutation operation is shown as figure 8. In which, child is generated by replacing node A and its sub-tree with new sub-tree. Figure 9 is the images correspond to the imaginary part of f(z) in figure 8.

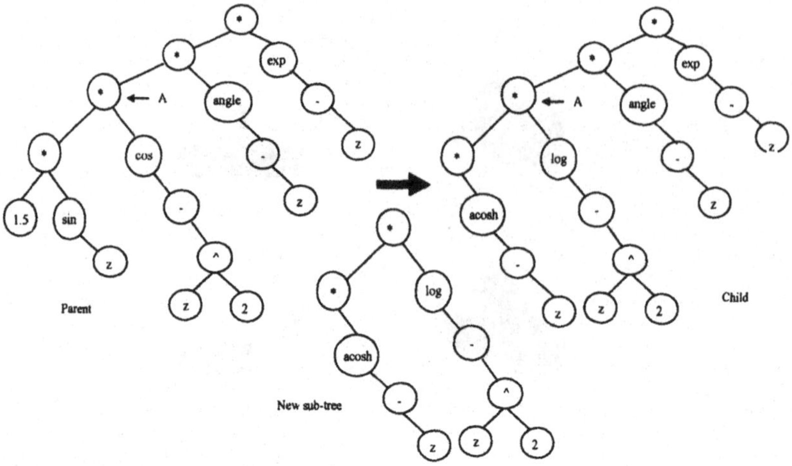


**Fig. 8.** The results of one mutation operation

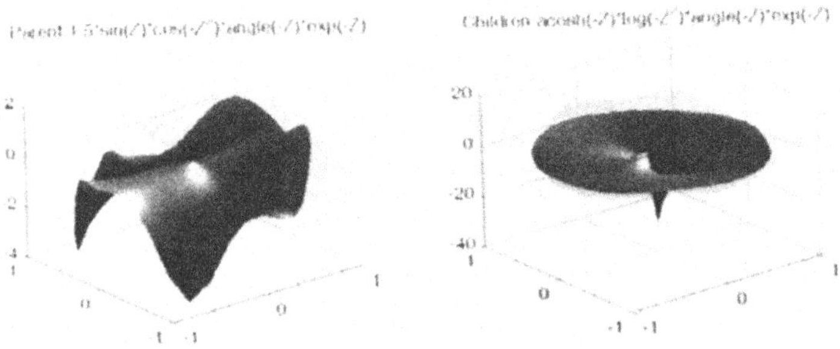

**Fig. 9.** The images correspond to the imaginary part of f(z) in figure 8

Step 5: If the procedure doesn't stopped by the designer, go to step 2.

This process of selection and crossover, with infrequent mutation, continues for several generations until it is stopped by the designers. Then the detail design will be done by designers with human wisdom.

The generated images are handled by designers using computer operations, such as rotating, cutting, lighting, and coloring and so on. The interactive user interface can be seen in figure 10. Then, we can get some artistic flower images as shown in figure 11.

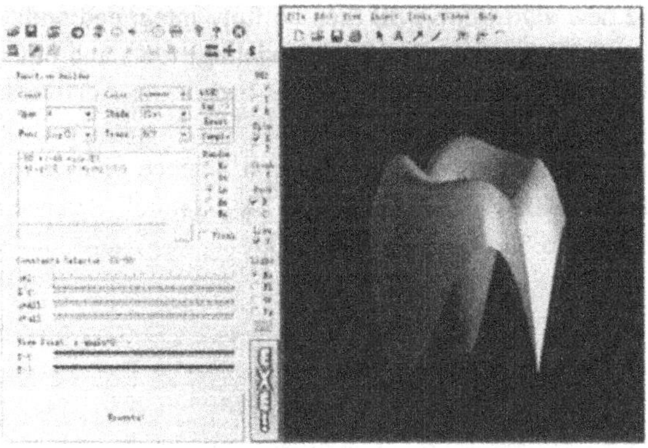

**Fig. 10.** The interactive user interface

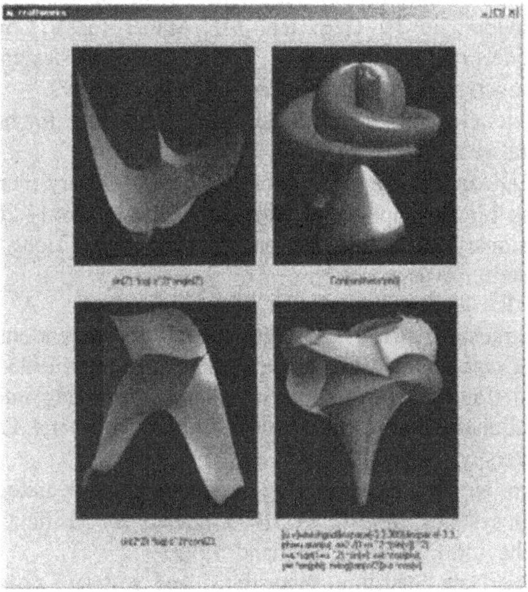

**Fig. 11.** Some artistic flowers generated by complex function expressions

# 5 Conclusions

Generative design is still in its infancy. There is still much work to be done before the generative design system can be in practice. Our current work is to use the multi-agent architecture as an integrated knowledge-based system to implement a number of generative design techniques including genetic algorithms and fractal geometry. These new algorithms will then be fully integrated with a selected set of 2D (sketching) and 3D (surface and solid modeling) tools and other design support systems. This integrated system is intended for supporting creative design in a visual environment [12].

This project is founded by National Natural Science Foundation of China (No. 69975010, No. 60374054) and supported by Natural Science Foundation of Shandong Province (No. Z2004G02, Z2006G09).

## References

1. C. Soddu. Argenia, naturality in industrial design, In: Proc. 3rd International Conference on Computer-Aided Industrial Design and Conceptual Design, CAID&CD'2000, International Academic Publisher, World Publisher Corporation, December 2000.
2. R. Saunders, J.S. Gero. Artificial creativity: A synthetic approach to the study of creative behaviour, in JS Gero and ML Maher (eds), Computational and Cognitive Models of Creative Design V, Key Centre of Design Computing and Cognition, University of Sydney, Sydney, 2001,pp. 113-139.
3. J. H. Frazer etc. Generative and evolutionary techniques for building envelope design .Generative Art 2002.
4. J. S. Gero, V. Kazakov. An exploration-based evolutionary model of generative design process. Microcomputers in Civil Engineering 1996, 11:209-216.
5. C. Soddu,Visionary variations, personal exhibitions at Hong Kong Museum, Visual Art Centre, 2002.
6. I. DeFalco, R. DelBalio,A. DellaCioppa,E Tarantino, A parallel genetic algorithm for transonic airfoil optimisation, In: Proc. International Conference on Evolutionary Computation 1995, Perth, Western Australia, 1995.
7. Y. Nakanishi, Capturing preference into a function using interactions with a manual evolutionary design aid system. In: Proc. Genetic Programming '96, Stanford University, 1996.
8. M. Pontecorvo. N. Elzenga, Exploring designer/consumer dialog in evolutionary product design systems, In: Proc. AISB'99, Society for Artificial Intelligence and Simulated Behavior, Edinburgh, Scotland, 1999.
9. A. S. Thomas, Data Structure, algorithms, and software principles, Addison-Wesley Publishing Company, inc.U.S.A. 1994.
10. B. Mckay, M. J. Willis, G. W. Barton, Using a ree structured genetic algorithm to perform symbolic regression, In Proceedings of the First IEE/IEEE International Conference on Genetic Algorithm in Engineering Systems: Innovations and Applications, Halifax Hall, University of Sheffield, UK. 1995, 487-498.
11. H. Liu, M. X. Tang, J. H. Frazer. Supporting dynamic management in a multi-agent collaborative design system. International Journal of Advances in Engineering Software, 2004, 35(8-9): 493-502.
12. H. Liu, M. X. Tang, J. H. Frazer, Supporting creative design in a visual evolutionary computing environment. International Journal of Advances in Engineering Software, 2004, 35(5): 261-271.

# Supporting the Sytematization of Early-Stage-Innovation by Means of Collaborative Working Environments

Alexander Hesmer; Karl. A. Hribernik; Jannicke Baalsrud Hauge; Klaus-Dieter Thoben
BIBA – Bremen Institute of Industrial Technology and Applied Work Science at the University of Bremen, Hochschulring 20, 28359 Bremen, Germany
{hes, hri, baa, tho}@biba.uni-bremen.de

**Abstract.** Research in the area of the early-stage of innovation concentrates on non-linear innovation environments constituted by the nature of the "fuzzy front end" of innovations in which there are no well-defined problems or goals at that point in time [1,2]. Early-stage-innovation requirements are the general applicability and the support of iterations within the software tools to be developed within future collaborative working environments (CWE). The research presented in this paper focuses on innovators' every day work and the related needs in todays and future work environments to provide a highly flexible software solution supportive to the early-stage-innovation. The adaptability of the software tools to – which depends on the fulfillment of the users requirements - will be achieved by supporting the real-life work routines of innovation workers and teams; be they co-located or dislocated. In the actor-network theory [3] early-stage-innovation is seen as a social process. Therefore the participation of individuals will be encouraged by the usage of game dynamics to supporting idea generation related workflows. To equalize the dependences of people working together in one place, time zone and personal relationship a database of knowledge and object representations will be implemented in the CWE. The CWE tools support and guide innovators to get connected to the right people, produce ideas based on explored knowledge and evaluate them to achieve the goal of developing successful innovations. The approach presented in the proposed paper is basing on the work carried out by the European funded research project Laboranova.

## 1 Introduction

Nowadays a boom in innovation is taking place in society. Innovation is the key to the advantage of western economies against its competitors from today's success-

---

*Please use the following format when citing this chapter:*

Hesmer, A., Hribernik, K. A., Hauge, J. B., Thoben, K.-D., 2007, in IFIP International Federation for Information Processing, Volume 250, Trends in Computer Aided Innovation, ed. León-Rovira, N., (Boston: Springer), pp. 135-144.

ful economies and the upcoming competitors from Asia and other emerging countries. In order to achieve continuous strategic innovation and thus create persistent competitive advantage, organizations and companies need to increase their capacity for carrying out open-ended and nonlinear problem solving involving a wide participation of people in knowledge-rich environments. Companies are well aware of this issue and have implemented strong innovation processes which are often represented by the stage-gate model. Most of these innovation processes have the black box in the beginning of the process called idea generation in common. As the output of an innovation process depends on the given input, and therefore is one of the main success factors, the early-stage of innovation is worth looking at and thereby support the generation of quantitatively and qualitatively better input to the innovation process.

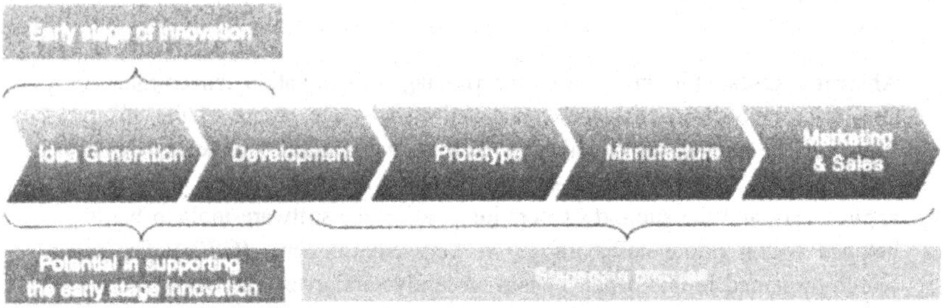


Fig. 1. The Innovation Process [4]

## 2 Theoretical Background of Early-Stage Innovation

*Innovation in Theory*

Innovation can be understood as the process where something new and valuable to a society is created and an economic advantage can be taken. The definition of Schumpeter [5] constitutes innovation as a new combination of resources [5]. Within all definitions of innovation there is a consensus on the following points:

- Innovation is not identical to invention, the main difference being that innovation covers the whole process from a new idea to a realized product or process available to potential users or customers,
- Innovation is a result of a number of intended actions, and not just the spontaneous nearly evolutionary development of new products and processes,
- Basically innovation is related to change and the emergence of something new – and not only new but in some respects better (thus innovation is often seen as a form of problem-solving), Innovation can be based on adaptation and evolution, but is not identical although a series of spontaneous adaptations can appear as an innovation.

Latest developments in literature present a view from a sociological perspective upon innovation and the change from a linear process — from research to innovation — to a user centric approach where both the technological research and the sociological aspects of innovation are addressed equally. Additionally strategic management and innovation are no longer perceived as a linear but as a parallel development. Innovation can be seen as learning/knowledge process within a community of practice.

*Ideation in Theory*

The ideation process, also called the "fuzzy front of innovation" is described to be the process of discovering what to make, for whom, understand why to make it and define the success criteria including the development of insights for answering these strategic questions [6]. Ideation as part of the overall innovation process is defines by Vaghefi [7] as the "ability one has to conceive, or recognize through the act of insight, useful ideas". Therefore ideation represents a process with ideas as a result. The "fuzzy front end" of innovation is one main factor to the success of an innovation project. Idea generation is often seen as the inspiration or intuition of an individual [8]. But idea generation can also be seen as outcome out of a work process not only related to an individual but to a group of people working together in a network. Innovation takes place more and more in distributed teams where collaborative working environments support the communication between workers and provide "shared access to contents and allowing distributed actors to seamlessly work together towards common goals" [9].

*Design as the Science related to Ideation*

The scientific discipline related to the ideation process is the theory of design methodology. According to Charles Eames [10] design is described as "a plan for arranging elements in such a way as to best accomplish a particular purpose". Design is seen as a discipline dealing with the early-stage of innovation.

Wolfgang Jonas [11] claims that today's efforts heading for the development of planning practices and methodological approaches without having the pretence of planning everything complete. This is consistent with Akin's [12] theory that states that "no quantifiable model is complex enough to represent the real-life complexities of the design process".

One reason for this issue is that one of the specific aspects of the working process of designers is the constant generation of new task goals and redefinition of task constrains [12] . In relation to information technology (IT) support Rahe [13] states that the problem with the most planning instruments is the inattention on the fact that during a development process new knowledge is achieved that changes the project. This underlines the thesis of non-linearity in the early-stage-innovation. In this context a proposal from Schön [14] comes into play who states to search for an epistemology of practice implicit in the worker, intuitive proceedings. The user centric approach is becoming more and more important to organizations.

Moreover organisations can utilise their internal resources better by making the implicit knowledge of employees available for the organisation. Approaches in this field are the so called "skunk work" where 10 - 20 % of the working time is dedicated to employees own projects.

*Current Support of Ideation*

Because of its fuzzy nature, where details and even goals are not defined exactly the early-stage-innovation can not take place in a linear process. Iterations are the nature of the related workflows. Traditional project management is all about linearity. But in the early-stages of innovation, one rarely has a well-defined problem, and so iterations between problem, solution and possibilities are needed [1].

Existing Collaborative Working Environments (CWEs) [15,16] mainly focus on supporting traditional working paradigms of linear workflows by providing IT-based platforms for planning, scheduling and executing tasks [17]. When comparing the theoretical foundation and the state-of-the-art in organizational innovation methods and approaches it seems to be obvious that today's software tools often represent a conversion of one methodology or the integration of a couple of these but they won't fulfill the requirements of future within a more open innovation culture. The usage of these tools is very seldom [18].

*Requirements for successful Ideation Support Tools*

However, in order to achieve continuous strategic innovation and thus create persistent competitive advantage, organizations need to increase their capacity for carrying out open-ended and nonlinear problem solving involving a wide participation of people in knowledge-rich environments. This must be supported by the next generation CWE's, which in turn, requires new paradigms for managing the knowledge transfer, the social dynamics, and the decision processes involved in the front-end of innovation.

With respect to this the actual research in the field of early-stage-innovation focuses on the real requirements of innovators in distributed working environments and the solving of the occurring challenges.

## 3 Research Approach

Looking at how systematization in early-stage-innovation takes place in companies one recognizes that workflows are based on an individual level or at least on group dynamic level [14]. The acceptance and usage of methods and tools in this field is very weak. Specific tools to support companies' processes in the early-stage of innovation are used seldom [17]. This is caused on an individual level by the evaluation of one regarding the benefit on the one hand side and the usability on the other hand side are parameters for the usage of tools [17]. If the effort of learning and using a tool and align with that change ones work process is higher than the expected gains to use it a tool will not be used [18].

To build a successful software solution that will be adapted and used in companies and networks of innovators one needs to build upon every day requirements and workflows people are already used to and are not willing to change. Therefore the all day work of ideators — knowledge workers in the field of innovation — is evaluated in research.

The ongoing work is based on the knowledge about the state-of-art in innovation and design theory and insides of best practice analysis of innovative companies.

To gather the information needed ideators and groups working together are observed and interviews are accomplished. Within the observation this individuals and groups are accompanied through their daily business. All activities are monitored and captured and put into the context of the actual task and workflow. In relation to this their organization of data and information – digital and physical representation - is observed. The usage of physical elements and IT tools is investigated. Further data is gathered by interviewing innovative workers lead by a questionnaire.

Based on this information workflows and routines can be identified and represented. Conclusively generic elements to support within an innovation environment and software tools supporting the early-stage-innovation can be extracted.

# 4 An Approach for the Support of Early-Stage Innovation

Results of the observation and interviews clearly show that creating and developing ideas is based on iterative routines of representing an idea, sharing it with others, getting feedback and communicating about the object (the representation). The approach presented supports the creation and development of ideas, viewing ideation as a working process rather than moments of divine inspiration. It will identify explicit routines for team-based ideation work, together with a technological infrastructure that allows for communication about, and experimentation with more or less finished ideas, early stage innovations and concepts not yet realized.

Representations of ideas can be e.g. sketches, renderings or maps. Work routines show that individual ideators represent their ideas in an "easy to access" way, meaning that CAD or rendering software is used in a basic way, more often ideas are sketched or presented in PowerPoint. The interviewees stated that the rational for using Microsoft (MS) PowerPoint is based on the one hand on the generic usage and on the other hand the exchange with others because of its status as a de facto standard of the product. The representation is distributed to stakeholders by mail for getting feedback in general, comments, further ideas, and the development of the original idea.

**Example of an Idea Development Routine**

The initial moment is the occurrence of an idea. This is not further specified. Within the time of one to eight hours the idea is represented as a sketch rendering or text. There might be variations of the idea but not an entirely different concept. Pictures are pasted in common media programs like MS PowerPoint or MS Word.

The document is sent out by e-mail to the recipients who have an interest in the idea. Usually they are well known. The reply by email occurs during two days otherwise there will be no reply at all after that. Alternatively feedback can be gathered by phone. Feedback is usually given in an unstructured way.

The feedback is extracted from the individual sources (text, comments to the pictures/text, phone calls) and than gathered. The feedback is then used to transform the original idea.

With this developed idea as the objective the routine starts again. The overall time frame for the described routine is about three to four days in total.

Core of this routine is the representation thereof its exchange with others. The interviewed person states that he stops thinking about how to developing the idea when not interacting with others.

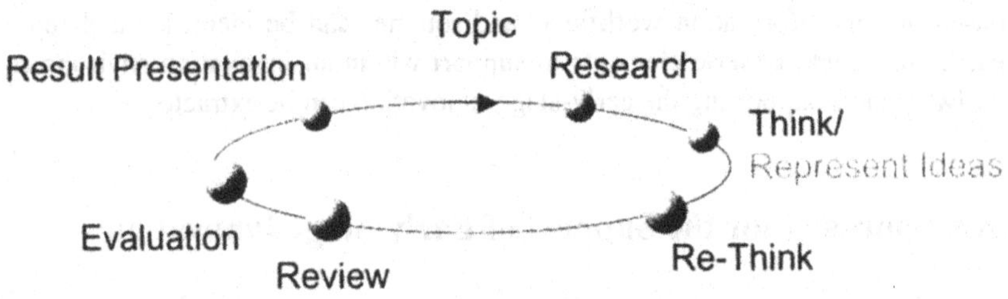


**Fig. 2.** Example of an Idea Development Routine

**Example Idea Generation Routine Group Perspective**

Within the observed group the first step is to show the discussion topic. It is visualized to a whiteboard or flipchart (large representation plane). The topic is discussed within the group to achieve a common understanding (verbal).

To generate ideas, brainstorming takes place supported by "Post Its" which are randomly placed on the representation plane. Ideas are affected by former thoughts and experience of the participants.

The next step is the structuring of the ideas to higher aggregation levels. This is done by discussing the ideas and finding group during that discussion. Within the discussion the ideas are usually evaluated on best guess basis. The ideas are clustered on the wall. For this step lots of space is needed to develop clear clusters. The possibility to edit the visualization with e.g. connecting lines with a marker is given when using a whiteboard.

The representation is captured by taking a photograph.

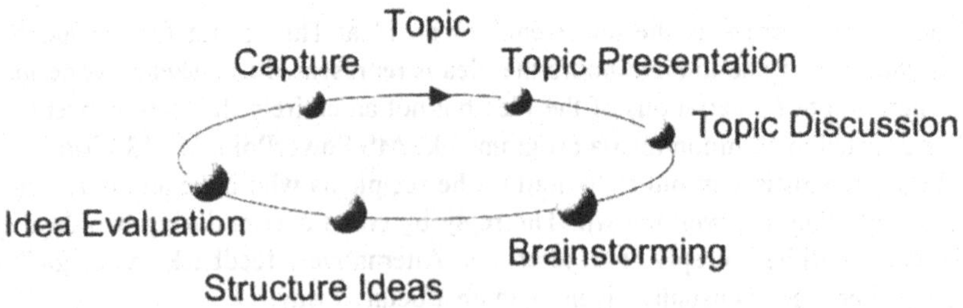


**Fig. 3.** Example of an Idea Generation Routine by a Group

Common to both of the examples is the importance of representing the idea one has and the exchange. People need feedback to further develop their idea. The externalization of ideas is mostly supported by generic media, like sketching on paper or - by means of IT support - the usage of MS PowerPoint.

People who are not physically work together or don't have the opportunity to have a meeting with other stakeholders regarding the idea send them out by email to provide the represented idea to them.

The represented object and the communication regarding it take place in several proprietary software solutions which might cause the loss of information or data inconsistency.

By means of transferring this knowledge in an innovation environment which merges the representation and the communication regarding the represented object to support innovators with their daily work a successful software solution can be developed.

The network of people working in the field of idea generation are neither not necessarily located in one place, even not in one company, nor do they work in one time zone. New connection mechanisms need to be developed and implemented to bring the right people together who share a specific interest. As much as participating in the idea generation process the motivation of individuals is key to success.

But creating connections is not only a matter of bringing the right people together. They can be instantiated between ideas to describe the intellectual lineage of an idea (e.g. where it is coming from) or to keep together related ideas. Connections transform a collection of ideas into a structure that can be browsed and filtered according to innovators' needs.

Even if the right individuals and the right ideas have been brought together and a quantity of ideas are generated in the provided environment, there are differences in the quality of ideas and the chance of success later within the market launch. In order to select the ideas which have the greatest chance of becoming a successful innovation evaluation is necessary. To achieve this goal the "intelligence of many" will be used by implementing a prediction market into the innovation environment.

## 5 Collaborative Working Environments

As early-stage-innovation takes place in diverse environments (e.g. on individual level, in groups, SME, companies, open innovation, Living Labs [19]) the CWE tools which will be developed need to be based upon a scaleable system. The tools can both be used integrated as well as single solutions related the needs of an individual, a group or company working in the field of early-stage-innovation (e.g. idea generation tools, evaluation tools).

The main modules within the innovation environment in Laboranova will be:

*Idea Database*

Ideas as outcome of early-stage innovation and also as working objects will be handled within an idea database. Implementations of editing and evaluation tools will lead to developed ideas. The idea generation will be supported by idea generation games for example based on the method of "Reframing the Question". Additional games in the field of idea generation will be developed.

*Idea Evaluation*

The Laboranova idea evaluation space aims at developing a group decision support system supporting decision makers in taking knowledgeable decisions. To do so, the idea evaluation space should support effective collaboration between decision makers in order to build consensus and collaboratively select the best ideas. Moreover, the Laboranova idea evaluation space will provide mechanisms for providing decision makers with the necessary knowledge to select the best ideas. This will be supported by a prediction market.

*Connection Space*

Laboranova will provide a user database with user profiles for describeing the experts interests and abilities. The proposed User Profile for use in Laboranova is a modular one. It consists of standard modules which are fixed (always present in a user's profile) and extensible modules which are modules that can be dynamically added by the system to the profile. These latter can be displayed on the User's Profile (visible) or hidden (invisible - only accessible by the system). The connection with experts one user might not know will be supported by connection games.

*User engagement by game dynamics*

For generating ideas games will be used for shorter, specific work routines. The game approaches for idea generation will be designed from the assumption that (good) ideas do not just come into existence but involve some analytical and explorative work. The objective of these ideation games is to promote and support innovative work. Most games available for companies are simulations mainly focusing on learning or team-building. For games to be used in ideation, and not just in training people in ideation work, the game should provide insights as well as make the participants able to act on these insights by coming up with ideas for new products, services and strategies.

The notion "game" is an ambiguous term – for some it signals energy, entertainment and creativity, while for others it signals a lack of seriousness and value. This implies that the diffusion and implementation of innovation games should focus on the productive side of the process. The message should be clear that while being a game the process is still work and should be taken seriously.

The follow-up process should be an integral part of the design of a game. Knowledge developed during the game should be documented and presented to the participants. Competences developed should be followed-up with action plans for further development, implementation and integration into ordinary practices. If the game is supposed to create input to decision processes in the organization, feedback to the participants about how the feedback should be communicated should also be part of the game's results.

## 6 Conclusions

To support early-stage innovation in distributed teams CWE need to be developed which support non-linear work processes. These iterative processes will be supported by the innovation environment in a way that does not change the habits and routines of people working in the field of innovation but provides tools and methods to them which augment the efficiency of their way of working. Important to this concept is the support of object related communication. It can be seen in the routine examples idea development is based on the representation of the idea, exchange of its representations and gathering feedback and get input to further develop the idea.

An IT based innovation environment with rated ideas on several development levels will support innovation workers with presenting and communicating their ideas to stakeholders, developing their ideas further, finding related ideas and people and will be the backbone to enhance companies ability to generate successful innovations.

Creating connections is not only a matter of bringing the right people together. Connections provide the backbone for ideation. They can be created between ideas to describe the intellectual lineage of an idea or to keep related ideas together. Connections transform a collection of ideas into a structure that can be browsed and filtered according to innovators' needs.

Even if the right individuals and the right ideas have been brought together and a quantity of ideas are generated in the provided environment, there are differences in the quality of ideas and the chance of success later within the market launch. In order to select the ideas which have the greatest chance of becoming a successful innovation evaluation is necessary. To achieve this goal the "intelligence of many" will be used by implementing a prediction market into the innovation environment.

The objective of the game is to make the work routine of generating ideas more effective through the use of games. The outcome of the game intended to be initial ideas but could also be broader and imply "options", e.g. ideas for solutions for specific problems. However, with focus on the fuzzy front end of innovation, the very early part of a project when the idea has not been found and the criteria for selecting a good idea are unclear and it is not sure that the idea will lead to a new product. The challenge of introducing and developing a game is that it should be possible to use it in a productive way, i.e it should be included in the work flow in generate

The overall goal is to provide CWE tools related to early-stage-innovation collected in an innovation environment which can be used easily; where innovators see the advantage of usage and by using it enhance the environment in its quality.

## References

1. Simon, H. (1973): The Structure of Ill-structured Problems, originally published in Artificial Tony J. Watson. Rhetoric, Discourse and Argument in Organizational Sense Making: a Reflexive Tale. Organizational Studies, 16(5):805–821, 1995. .

2. Bayazit, N. (2004): Investigating Design: A Review of Forty Years of Design Research, essay in Design Issues, Volume 20, Number 1
3. MIT, Allan, (January 7, 2007); http://esd.mit.edu/HeadLine/allen030106/allen030106.htm
4. R. Rothwell, Successful industrial innovation: critical factors for the 1990s,
1. R&D Management 22 (3), 221–240 (1992).
5. J.A. Schumpeter, Theorie der wirtschaftlichen Entwicklung, (Dunker&Humblot, Berlin, 1952).
6. D, Rhea, Bringing Clarity to the "Fuzzy Front End" - A predictable Process for Innovation, Design Research (The MIT Press, Cambridge, 2005)
7. M.R., Vaghefi, and A.B. Huellmantel, Strategic Management for the XX Century (Boca Ranton, 1998).
8. W. Weisberg, Creativity, Beyond the Myth of Genius (Freeman, New York, 1993).
9. K.A. Hribernik, A Set of High-level Objectives, Definitions and Concepts, Brain Bridges IST-015982 (2005).
10. Design Within Reach, Design Notes (February 15, 2007): http://www.dwr.com/images/newsletter/20061010_salads/index.html.
11. B. Bauer, Design & Methoden, in: Design Report 11/06 (Blue C. Verlag GmbH, Leinfelden-Echterdingen, 2006).
12. N. Cross, Designerly Ways of Knowing: Design Discipline Versus Design Science, paper prepared for the Design+Research Symposium held at the Politecnico di Milano, Italy, May 2000 (2001: MIT Online)
13. P. Klünder, Planbarer Brückenschlag, in: Design Report 11/06 (Blue C. Verlag GmbH, Leinfelden-Echterdingen, 2006).
14. N. Cross, Developments in Design Methodology (John Wiley & Sons, Chichester, 1984).
15. K.A. Hribernik, K.-D. Thoben, M. Nilsson, Technological Challenges to the Research and Development of Collaborative Working Environments in: Encyclopedia of E-Collaboration (Idea Group Reference, 2007).
16. K.A. Hribernik, K.-D. Thoben, M. Nilsson, Collaborative Working Environments. in: Encyclopedia of E-Collaboration (Idea Group Reference, 2007).
17. Nova-Net Konsortium, Nutzung von Internet und Intranet für die Entwicklung neuer Produkte und Dienstleistungen (Fraunhofer IEB Verlag, Stuttgart, 2006).
18. C. Frey, Mind Mapping Software Survey, Innovation Tools, http://www.innovationtools.com/survey/index.asp, Accessed: 13.02.2007
19. M. Eriksson, V.-P. Niitamo, S. Kulkki, K A. Hribernik, Living Labs as a Multi-Contextual R&D Methodology. in: 12th International Conference on Concurrent Enterprising (ICE 2006). Proceedings. Milan, Italy 2006

# Constraint based modelling as a mean to link dialectical thinking and corporate data. Application to the Design of Experiments.

Thomas Eltzer, Roland DeGuio
INSA Graduate School of Science and Technology - LGeCO
24 Boulevard de la Victoire, 67084Strasbourg Cedex FRANCE.
eltzer@yahoo.fr, roland.deguio@insa-strasbourg.fr

**Abstract.** Problem solving is a key activity for any innovative company. As the dialectical approach has shown its efficiency in problem solving, the starting point of our work is the use of contradiction in problem formulation stage. In this article, we analyse the possibility to use information brought by a Design of Experiments (DoE) to formulate a contradiction. Constraint Satisfaction Problem is employed to present information brought by the DoE. To use the dialectical approach, we define the concept of the Generalised Contradiction (GC). In the framework of DoE data, we show how to formulate a GC, and propose an algorithm able to identify OTSM-TRIZ System of Contradictions. Results and perspectives are discussed.

## 1 Introduction

Any manufacturing company often faces the need to solve problems: quality problems, industrialisation problems, design activity, etc. In this article we adress the domain of innovation problems. Solving an innovation problem means obtaining a product description detailed enough to manufacture it, [1]. Of course, this product should have precise properties (some of which are known at the beginning of the solving process). A part of innovation problems can be solved using an optimisation approach: the parameters of the product remain the same and one searches their best possible values, [2]. However, some problems cannot be solved by such an approach: the set of required properties returns an empty solution space. This means that the product has to be completely rethought, because no satisfying solution will be found by simply adjusting the parameters values. The product architecture has to be redefined. For such a solutionless problem, optimisation techniques are useless: another solving approach has to be proposed to the designer. Any problem solving process contains at least two phases: problem formulation and problem resolution,

---

*Please use the following format when citing this chapter:*

Eltzer, T., DeGuio, R., 2007, in IFIP International Federation for Information Processing, Volume 250, Trends in Computer Aided Innovation, ed. León-Rovira, N., (Boston: Springer), pp. 145-155.

[3]. In this article, we focus on problem formulation and develop a new framework to formulate the problem of solutionless situations.

The dialectical approach has been developed by famous philosophers: Plato [4], Hegel, Marx, Mao Tse-Tung, [5]. The dialectical way to adress a problem has the three following bases: a thesis, its opposed antithesis, and the solution in which these two opposites can coexist. Thus, the core of the dialectical approach is the identification of a contradiction between a thesis and an antithesis. The dialectical way of formulating a problem is the formulation of such a contradition. Dialectic-based problem solving approaches have shown a significant efficiency in, among others, the following domains: information system design [6], manufacturing process design [7] or inventive technical problems [8]. In the third domain, a complete theory has been developed: TRIZ. Its development has been initiated by G. Altshuller. One of the main ideas is to use contradictions as a way to formulate problems. More than 50 years after the first publication on this problem solving approach, some evolutions have been proposed, among others: General Theory of Advanced Thinking (russian acronym OTSM), Theory of Development of a Strong Creative Personality (TRTL), Theory of the Evolution of Technological Systems (TRTS). It is in the border of OTSM-TRIZ that the idea of contradiction is the most clearly defined, [9]. The proposed System of Contradictions is inspired by the dialectical approach and artificial intelligence models. A schema representing the System of Contradictions (SC) is given Fig. 1. Facts concerning a classical laptop are used to illustrate the SC.

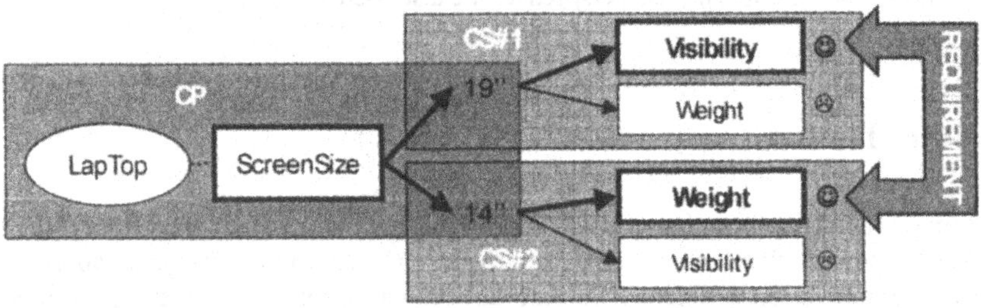


**Fig. 1.** System of Contradictions, applied to a laptop

The first component of a SC is the "requirement": two measured parameters, of which the value should belong to a predefined domain. The requirement describes the objective of the problem solving. The laptop weight and information visibility should be both satisfying. Weight should be low and visibility should be high. The second component of a SC is the "contradiction of a parameter": a parameter X should be assigned two different values. The screensize should be both 19'' and 14''. The remaining components of a SC are two "contradictions of a system": if one of the two possible values is assigned to X, one measured parameter is satisfying, and the second one is not. CS#1: if the laptop screensize is 19'', then visibility is good but weight is bad. CS#2: if the size is 14'', then weight is good, but visibility is bad. Problem solving principles have been built for the SC: 13 principles to solve a contradiction of a system, 6 principles to solve a contradiction of a parameter [10]. Thus, we will investigate the possibility to use the SC model in a complex set of information.

A SC is formulated between two pieces of information. For innovative design problems, the first available information source is the expert designer: a contradiction will be proposed on the basis of what the expert designer says. However, this information source has, among others, the following disadvantages: (1) it is difficult to extract tacit knowledge (2) an expert does not want to share everything (3) experts can be wrong (4) time required to gather enough and reliable information can be very long. Nowadays, other sources of informations are available especially because companies develop their own knowledge management activites, [11]. Part of our research activity investigates the possibility to use alternative information sources in the problem formulation phase, to complement or rectify expert knowledge. In this article, we focus on how to extract contradictions and SC out of a Design of Experiments.

## 2 Design of Experiments

The tools and techniques existing in the border of DoE can be grouped in two families which are modelling and optimisation. The first family of tools assists the pragmatic construction of mathematical models describing a measurable phenomenon, [12]. Parameters concerned by a DoE are either measurable or controlled. The controlled parameters are usually noted X (the value is chosen) and the measured parameter is usually noted Y (its value is only observed). Usually, the measured parameter describes a portion of the product requirement. A mathematical model is built on the basis of the set of achieved experiments. Experimental measures are usually listed in a chart (Table 1). One of the objectives of DoE tools is to obtain a realistic model with the minimum number of experiments. Robustness of the obtained model can be valuated using statistical calculations. The number of requested experiments can be reduced using Tagushi's method [13], fractionnary plan technique, or by removing some parameters based on the experts opinion. The second family of DoE techniques concerns the exploitation of the obtained mathematical model, [14]. The major kind of exploitation is the determination of requested controlled parameters values: given a required value of the measured variable, the mathematical model is used to find requested controlled parameters values. Different optimisation algorithms exist, but as we said in the introduction, we do address in this paper problems that cannot be solved by optimisation as no solution do exist within the given model. In this article, we consider that the domain of each controlled parameter is continuous. The value chosen during an experiment is one out of a continuous domain.

Let's consider as an example that will be used along the paper, a problem with two controlled parameters $X_1$ and $X_2$, influencing three variables $Y_1$, $Y_2$ and $Y_3$ simultaneously measured. The values of $X_1$ and $X_2$ must be chosen within the range [-1,1]. The requirement is that $Y_1$, $Y_2$ and $Y_3$ should be greater than 6. We are looking for pairs ($X_1$, $X_2$) that fit this requirement by performing a DoE, the result of which is given in Table 1. Four experiments have been done.

**Table 1.** Example of DoE result

| Exp n° | $X_1$ | $X_2$ | $Y_1$ | $Y_2$ | $Y_3$ |
|---|---|---|---|---|---|
| 1 | -1 | -1 | 10 | 10 | 0 |
| 2 | -1 | +1 | 10 | 0 | 10 |
| 3 | +1 | -1 | 0 | 10 | 10 |
| 4 | +1 | +1 | 0 | 0 | 0 |

The obtained mathematical model describing the effect of $X_1$ and $X_2$ on $Y_1$, $Y_2$ and $Y_3$ is respectively:

$$Y_1=f_1(X_1,X_2)=-5.X_1+5 \quad (1)$$

$$Y_2=f_2(X_1,X_2)=-5.X_2+5 \quad (2)$$

$$Y_3=f_3(X_1,X_2)=-5.X_1.X_2+5 \quad (3)$$

Fig. 2 provides a graphical representation of the three models of effects. $Y_1$, $Y_2$ and $Y_3$ are plotted versus both $X_1$ and $X_2$. In each of the three diagrams, the grey floor represents the possible values for $(X_1,X_2)$ and the grid represents the consequent value of the measured variable. The range of colors is linked to the value of the measured variable. The posed problem is solutionless with the table and the mathematical model. Indeed, whatever the value of $(X_1, X_2)$ within the allowed domain, there is at least one measured variable which does not satisfy the requirements. However, as each couple of measured variable can be simultaneously satisfied (for example, test n°1 shows this for $Y_1$ and $Y_2$), no SC among the parameters of the given models can be found in this example. In this paper, we do not investigate the possibility to change the problem definition to fit the SC pattern, as our purpose is to automatically extract contradictions from a given model.

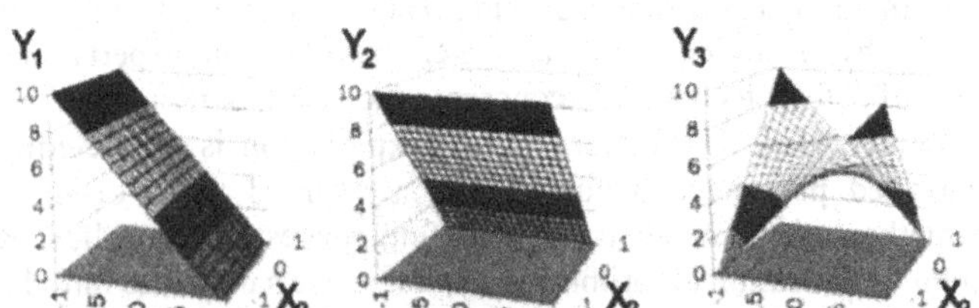


**Fig. 2.** Graphical representations of the relations among the five parameters

This simple case example shows that there is still a need for a problem formulation pattern, based on the dialectical approach, which could be used for any solutionless model. Table 2 lists our proposed criteria for this kind of problems.

**Table 2.** Criteria to evaluate the proposed problem model

| Criteria n° | Description | Comment |
|---|---|---|
| C1 | Existance of solving tools that can be applied on the problem decription | This is an important criterion, as the goal of problem formulation is not only description, but mainly resolution. |
| C2 | The problem formulation pattern describes a property that can be found in any problematic situation | This second criterion directly emerged from the case example which cannot be described using OTSM-TRIZ model |
| C3 | The problem formulation pattern describes a property that is absent of any solvable problem | The dialectical approach efficiency has only been shown in solutionless problems. |
| C4 | The problem is completely described with a single contradiction | If this is not true, the problem solver will have to select one of them, and, even if the selected contradiction is solved, the problem can remain unsolved. |
| C5 | The problem formulation pattern provides a synthetic description of the problem | In the other case, it might be difficult for humans to handle the description |

# 3 Constraint Satisfaction Problems

## 3.1 Definition

Montanari introduced the constraint satisfaction problem (CSP) model in [15]. Any CSP comprises:

- a set of n variables $V'=(V_1,..,V_n)$. A value of $V_k$ is noted $v_k$. Therefore, v' is defined as the tuple $(v_1,..,v_n)$;
- a set of domains $D'=(D_1,..,D_n)$ where $D_i$ is the finite set of possible values for $V_i$;
- a set of constraints $C'=(C_1,..,C_j)$ between the existing variables. A constraint C on a subset of V' specifies the allowed combination of values for the variables.

A solution to a CSP is an assignment of a value from its domain to every variable, in such a way that all constraints are satisfied at once. A constraint is not satisfied when the assignment of the variables is not allowed by the constraint. A CSP is solvable if at least one solution exists. Two analyses can be done on a CSP: find one solution, or find all solutions. These two objectives gave birth to different solving algorithms, [16].

Therefore, controlled and measured parameters of a DoE simply become "variables" in a CSP. Controlled parameters are grouped in $X'=(X_1,..,X_m)$ and measured parameters are groupped in $Y'=(Y_1,..,Y_p)$. Of course, $V'=X' \cup Y'$. Domains of controlled parameters are grouped in $D'_X=(D_{X1},..D_{Xm})$ and domains of measured parameters are grouped in $D'_Y$. An assignment of a value to each control parameter is named $x'=(x_1,..,x_m) \in D'_X$. Concerning the measured variable $Y_k$, the domain $D_{Yk}$ is the set of satisfying values. In the case example, as $Y_1$, $Y_2$ and $Y_3$ should be not less than 6, $D_{Y1}=D_{Y2}=D_{Y3}=[6;+\infty[$. The solution is not satisfying if one of them is out of this domain. For each parameter $Y_k$, its assigned value $y_k$ should belong to $D_{Yk}$. Concerning the controlled variables, the domain is the complete range of possible values. In the case example, the domain for $X_1$ and $X_2$ is the same. $D_{X1}=D_{X2}=[-1;+1]$. At this stage, the CSP model representing the case example is defined by:

- the set of five variables. $V'=X'UY'=(X_1,X_2,Y_1,Y_2,Y_3)$. $X'=(X_1,X_2)$; $Y'=(Y_1,Y_2,Y_3)$;
- the set of domains: $D'=(D_{Y1}, D_{Y2}, D_{Y3}, D_{Y4}, D_{X1}, D_{X2})$.

### 3.2 Design of Experience and Constraints

In this section we show in which shape the result of a DoE can be found in a CSP. The DoE often ends by the definition of a mathematical model, linking the controlled parameters and one measured parameter. Thus, this mathematical model directly specifies the allowed combination of values for these parameters. In fact, the equation describing this mathematical model defines the constraint linking the controlled parameters and the measured variable. Therefore, as we have made the hypothesis that the values of the controlled parameters are one out of a continuous range, the set of achieved experiments allows the definition of a constraint in intension. This constraint links p variables, and only one of them is measured.

In our case example, as there are three measured variables, three constraints are defined: $C'=\{C_1,C_2,C_3\}$. $C_1$,$C_2$ and $C_3$ contraints are defined by Eq.1, Eq2 and Eq.3. The CSP of the case example is now completely defined by V', D' and C'.

## 4 The proposed framework: the Generalised Contradiction

### 4.1 Definition

Applying the dialectical approach on a solutionless problem comes to present it as an opposition. The very first detected opposition is between the required values of the set of measured parameter and the possible values of the set of controlled parameters. "The problem is unsolvable" means that there is no way to simultaneously create a satisfying value for each measured parameter. Each time the controlled parameters are tuned to satisfy $Y_i$, at least one other measured parameter $Y_{j,(i\neq j)}$ is not satisying. This means that the set of controlled parameter values able to satisfy $Y_i$ and the set of values able to satisfy $Y_j$ have no intersection. Let us define $S_{Yi}$ as the set of values of $X'=(X_1,..,X_m)\in D_X'$ which are able to guarantee that $Y_i$ is satisfying (Eq.1). It should be noted that if there is one, two or three controlled parameters, $S_{Yi}$ geometric interpretation is respectively a line, a surface, or a volume. We define the Generalised Contradiction as the property of Y' detailed in Eq.5: there is no common value to all $S_{Yi}$. This property is equivalent to "problem is unsolvable with the given model", as proved by the two following demonstations:

- if the property described by Eq.5 is wrong, then it means that it exists an assignment x' to X' which belongs to any $S_{Yi}$. This means that x' creates a satisfying value of any $Y_i$. This means that the problem is not solutionless: if X'=x', each $Y_i$ is satisfying. Hence, if Eq.5 is true, the problem is solutionless;

- if the problem is not solutionless, then it exists an assignment x' to X' able to create a satisfying value to each $Y_i$. Hence, by definition, x' belongs to any $S_{Y_i}$. Thus, the intersection is not empty. Hence, if the problem is solutionless, Eq.5 is true.

$$S_{Y_i} = \{x' / (y_i = f_i(x') \in D_{Y_i}) \; AND \; (x' \in D'_X)\} \tag{4}$$

$$\bigcap_{i/Y_i \in Y'} S_{Y_i} = \emptyset \tag{5}$$

Let us illustrate this property on the case example. As there are three measured parameters, there are three sets $S_{Y1}$, $S_{Y2}$ and $S_{Y3}$. As there are two controlled parameters, each of these domains has two dimensions, i.e it is a surface. Fig. 3(a) shows each of these three surfaces in the square of the possible values x' of $X'=(X_1,X_2)$. One can notice the three following facts: (1) $S_{Y1} \cap S_{Y2}$ is not an empty set (the bottom left square), it means that it is possible to reach a satisfying value for both $Y_2$ and $Y_3$ simultaneously; (2) neither $S_{Y1} \cap S_{Y3}$ nor $S_{Y2} \cap S_{Y3}$ are empty sets, which finally means that each couple of measured variable can be reached simultaneously; (3) $S_{Y1} \cap S_{Y2} \cap S_{Y3} = \emptyset$, which means that the problem is solutionless.

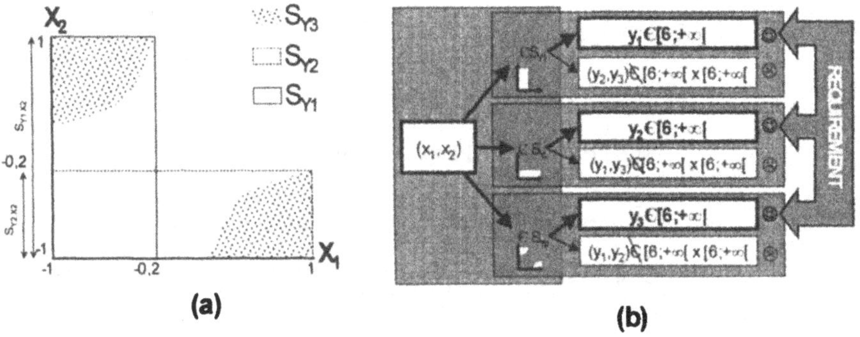


**Fig. 3.** (a) SY1 and SY3 (b) Possible GC graphical representation

In comparison, we can say that the System of Contradictions is concerned only by the intersection of a couple of domains $S_{Yi}$ and $S_{Yj}$ in one of their dimensions. Therefore, if we define $S_{Yi.Xk}$ as the set of values $x_k$ which can satisfy $Y_i$ (Eq6), we can say that the SC is based on the fact that $S_{Yi.Xk}$ and $S_{Yj.Xk}$ have no common value. However, we have shown that this condition is not equivalent to the absence of solution. In our case example, we have: $S_{Y1.X2}=[-1;1]$ and $S_{Y2.X2}=[-1;-0,2]$. The intersection is not empty because if $x_2 \in [-1;-0,2]$ we can satisfy both $Y_1$ and $Y_2$. However, the impossibity to simultaneously satisfy $Y_3$ is not seen.

$$S_{Y_i.X_k} = \{x_k / \exists (x_1,...,x_{k-1},x_{k+1},...,x_m)/((x_1,...,x_m) \in S_{Y_i})\} \tag{6}$$

In Fig. 3(b), we use the SC scheme to represent the GC of the case example:

- the requirement is that both $Y_1$, $Y_2$ and $Y_3$ are satisfying;

- $(x_1,x_2)$ shoud belong to $S_{Y1}$, $S_{Y2}$ and $S_{Y3}$;
- when $(x_1,x_2)$ belongs to $S_{Y1}$, $Y_1$ is satisfying but at least one variable of $(Y_2,Y_3)$ is unsatisfying; when $(x_1,x_2)$ belongs to $S_{Y3}$, $Y_3$ is satisfying but not $(Y_1,Y_2)$; when $(x_1,x_2)$ belongs to $S_{Y2}$ $Y_2$ is satisfying but not $(Y_1,Y_3)$.

### 4.2 Algorithm for getting SC

The following algorithm searches possible SC. The first step consists in calculating each $S_{Yi}$, then the i2-loop analyses each possible couple $(S_{Yi}, S_{Yj})$ and checks the intersection in each dimension (i3-loop). The output of this algorithm is the exhaustive list of SC. As two calculations have to be done for any triplet of $(X_k,Y_i,Y_j)$, the total number of calculation is $Card(Y')+2.Card(X').2^{Card(Y')}$.

```
FOR i1 = 1 TO Card(Y')

Calculate S_Yi1 'Achieved with the DOE mathematical model

  FOR i2 = 1 to i1 - 1

      FOR i3 = 1 TO Card(X')

          Calculate S_Yi1.Xi3 and S_Yi2.Xi3

          IF S_Yi1.Xi3 ∩ S_Yi2.Xi3 = Ø THEN

            Store the following SC:

            CP is "X_i3 ∈ S_Yi1.Xi3 AND X_i3 ∈ S_Yi2.Xi3"

            CS#1 is "if X_i3 ∈ S_Yi1.Xi3, then Y_i1 is satisfying
            but not Y_i2"

            CS#2 is "if X_i3 ∈ S_Yi2.Xi3, then Y_i2 is satisfying
            but not Y_i1"

          END IF

      NEXT i3

  NEXT i2

NEXT i1
```

# 5 Discussion

First, we briefly compare the strong and weak points of our contribution and of the System of Contradictions of OTSM-TRIZ according to the five criteria of Table 2.

C1: different solving principles have been adapted from TRIZ to OTSM-TRIZ. We have not yet developed the corresponding solving mechanisms for the GC framework. C2: the simple case example we developed is a solutionless problem. However, the SC could not describe it. We demonstrated that the GC basic property (Eq5) fits this criterion. C3: this is true for both the SC and the GC. C4: the System of Contradictions is based on the property of a triplet ($X_k$, $Y_i$, $Y_j$): $S_{Yi.Xk} \cap S_{Yj.Xk} = \emptyset$. In problems concerned by many measured and controlled parameters, this property can be found for more than one triplet, [17], [18]. This raises the following questions: which one should be chosen? how to navigate in and manage many contradictions? The GC is based on a property of Y' (defined in Eq5). This property is unique, as, for a given problem, there is a single Y'. C5: the System of Contradictions is very synthetic, as there are only three parameters. However, one can miss a lot of useful information. Describing a GC requires to list each $S_{Yi}$, which can be less synthetic. We can consider the GC is a generalization of the SC along these five criteria.

In this paper, we have addressed the problem of solutionless problems and proposed the GC framework to support the formulation phase. In the borders of constraints based problem modelling, solutionless problems are named "overconstrained problems": the constraints are so numerous that no satisfying value assignment is allowed, [19]. To face such a solutionless situation, the main current principle is known as "constraint relaxation". Constraints are simply removed one by one to open the solution space. A lot of research work concerns constraints relaxation order. Two main drawbacks of constraint relaxation are (1) if the constraint represents a scientific law (this is the case for the DoE mathematical model), constraint relaxation has no sense and (2) the objectif may not be reached, ie the product won't have each required feature. Our research results bring the first brick to build a new solving strategy. This strategy will consist in, first, formulating a GC to describe the problem and, second, applying solving mechanisms to this model. Extending the TRIZ solving mechanisms at the level of our GC framework is part of our current research. The second contribution in this domain is that the GC can link optimisation design tasks and inventive design tasks. Up to now a design problem had to be treated either with optimisation techniques (Operations Research algorithms to search for a solution in a CSP, [2]), or with inventive approaches (like TRIZ, for simple inventive tasks, [8]). No single information support could be used for these two tasks. Thanks to the proposed GC framework, information can be formalised in a CSP model and treated with both optimisation algorithms and inventive procedures. Furthermore, as CSP modelling is very generic and can be applied to numerous domains, the GC pattern can be applied to analyse other sources of corporate data, [11].

## 6 Conclusion

In this article, we have addressed the description of solutionless problems using the dialectical philosophy. We have considered that Design of Experiments can be used as an information source. The General Contradiction has been proposed to bypass the fact that, in complex cases, the System of Contradictions of OTSM is not equivalent to the absence of a solution. The Generalised Contradiction describes the fact that satisfying one requirement degrades at least another one. The Generalised Contradiction considers intersections of sets which dimensions equals the number of controlled parameters. One of the key advantages of the GC is the possibility to represent any complicated problems with a dialectical approach.

The first research direction consists in proposing problem solving mechanisms, based on those developed developed in TRIZ and OTSM-TRIZ, to fit the GC framework. The second research direction concerns the description of the GC. The fact that the intersection of SYi is empty can be described in many different ways: one domain is empty, one (or more) couple of domains has no intersection, one (or more) triplet of domains has no intersection, etc. One research activity aims now at finding which combination of parameters and domains should be used to provide a comprehensive and synthetic description of the GC. The third research perspective is the development of a more efficient algorithm to provide a comprehensive GC description. The first possibility is to use already existing CSP algorithms to provide automatic contradiction identification. The key question to answer is the equivalence between contradictions and consistencies, as defined in CSP modelling. The second possibility is searching only those contradictions separating the unreachable desired system from the today best reachable ones.

## References


1. W. BEITZ and G. PAHL, Engineering design - A systematic approach (Springer, 1996).
2. F. Hillier and G. Lieberman, Introduction to Operations Research, (McGraw-Hill, 2005).
3. S. Easterbrook, E. Beck, J. Goodlet et al, A survey of empirical studies of conflict, in CSCW: Cooperation or conflict (Springer, 1993).
4. S. Scolnicov (traducer) Plato's Parmenides (University of California Press, LA, 2003)
5. M. Tsetung, On Contradiction, in Selected Works, 311-347 (Foreign Languages Press, Peking, 1967).
6. V. Goepp, F. Kiefer and F. Geiskopf, Design of information system architectures using a key-problem framework, Computers in Industry, 57(2), 189-200, (2006).
7. F. Geiskopf, Formalisation et exploitation des contraintes Produit/Process pour la conception de systèmes de production - application à l'Usinage Grande Vitesse, PhD thesis, (ULP, Strasbourg France, 2004).
8. G. Altshuller, Creativity as an exact science, (Gordon and Breach Publishers, 1988).
9. N. Khomenko, R. De Guio and D. Cavallucci, Enhancing ECN's abilities to adress inventive strategies using OTSM-TRIZ, Int. J. Collaborative Engineering, (unpublished).
10. N. Khomenko, Introduction to OTSM-TRIZ. (Lectures of Advanced Master of Innovative Design, INSA Strasbourg, 2005).
11. N. Bontis, N. Dragonetti, K. Jacobsen and G. Roos, The Knowledge Toolbox: A review of the tools available to measure and manage intangible resources, European Management Journal, 17(4), 391-402 (1999).
12. D. C. Montgomery, Design and Analysis of Experiments, (Wiley-Interscience, 2004)
13. R. K. Roy, Design of Experiments Using The Taguchi Approach: 16 Steps to Product and Process Improvement, (Wiley-Interscience, 2001)
14. G. B. Dantzig, Origins of the simplex method, in A history of scientific computing, 141-151 (Reading, MA, 1990).
15. Montanari, Network of constraints: fundamental properties and applications to picture processing, Information science, 7, 95-132 (1974).
16. V. Kumar, Algorithms for constraint-satisfaction problems: a survey, AI Magazine, 13(1), 32-44 (1992)

17. D. Cavallucci and N. Khomenko, From TRIZ to OTSM-TRIZ: Addressing complexity challenges in inventive design, Int. J. of Product Development (unpublished).
18. E. Freuder and R. Wallace, Partial Constraint Satisfaction, Artificial Intelligence **58**, 21-70, (1992)
19. T. Eltzer, D. Cavallucci, N. Khomenkho, P. Lutz, E. Caillaud, Problem Formulating for Inventive Design Application to Injection Molding Technology, in Advances in Design, Chapter 3.6, (Springer, 2006)

# Development of Standard Solutions CAI System with UML and XML

Bojun Yang, Jianhui Zhang, Runhua Tan, Yumei Tian, Jianhong Ma
Hebei University of Technology, Institute of Design for Innovation,
Dingzigu, Hongqiao District, Tianjin, 300130, P.R.China
ybj@hebut.edu.cn
WWW home page: http://www.triz.com.cn

**Abstract.** CAI software products based on TRIZ were successfully applied in superior firms in the world. The Standard Solutions CAI system is one of branch of CAI software. Standard Solutions is important tool for product design. Standard Solutions for innovation firstly models a technical or process problem by Substance-Field (Su-F) Analysis, then synthesizes and converts the problem to acquire a solution. This paper researched on Standard Solutions arithmetic which can be applicable in CAI software product based on Standard Solutions, and established the flow of software. During the development of CAI software product, static models built based on UML were used to construct the configuration of system. For the components of system, dynamic models were used to describe the behaviors of system components. The Standard Solutions CAI software was coded with VC++ language. The interfacial view of this software, which is friendly for users, was created by VC++ combines with XML+XSL. Standard Solutions have great help to innovation design, farther more, its CAI software system help designer adequately apply the theory of Standard Solutions.

**Key words:** TRIZ, Standard Solutions, Computer-Aided Innovation, UML, XML

## 1 Introduction

Competition is the main impetus of technology progresses, also the resource of survival and development for enterprises. The contents of competition are change with varied time. Nowadays, most enterprises focus their attention on product design except for product qualities, cost, and production process and so on. Various methodologies and methods of product design progress quiet great recently years, like TRIZ, AD, QFD and TOC. At the same time, information technology applied

*Please use the following format when citing this chapter:*

Yang, B., Zhang, J., Runhua, T., Tian, Y., Ma, J., 2007, in IFIP International Federation for Information Processing, Volume 250, Trends in Computer Aided Innovation, ed. León-Rovira, N., (Boston: Springer), pp. 157-165.

widely to product design stage, from CAD and FEA, to CAI. Information technology is fusing in design methods more and more.

In the former Soviet Union, TRIZ [1], the theory of Inventive Problem Solving, was developed by Genrikh Altshuler and his school beginning in 1946. After more than 60 years development and application, TRIZ was proved powerful for concept generation and innovation design. Standard Solutions is one TRIZ knowledge of many parts and kinds of TRIZ. Standard Solutions include 76 items, can be used to solve relatively common optimization problems, and more particular to solve level three inventive problems [2].

The aim of Computer Aided Innovation (CAI) is help product designer can use innovative theory more effective and get more doable product design scheme. CAI software have many kinds, but TRIZ-based software is mainstream and extensive. From 1991, a TRIZ-based software package which developed by the Invention Machine Corporation was commercially launched [3], many CAI software based TRIZ have been developed now, such as TRISolver, IWB, Goldfire and so on.

To using Standard Solutions, this paper research on develop a CAI software system based on Standard Solutions. The system is a module of InventionTool3.0.

## 2 Standard Solutions and Theirs Usage

### 2.1 Substance-Field

According to the process of inventive problem solving, TRIZ system constructed from the theoretical foundation, laws of evolution, the analytical tools, such as contradiction analysis and required function analysis, and knowledge base tools, such as 40 principles and effects database. Substance-Field analysis, simply, Su-F analysis, is one of analytical tools or Search engines, and Standard Solutions is knowledge base tool (Table 1).

**Table 1.** The knowledge to support inventive problem solving

| Knowledge | Search engine | Levels of solutions |
|---|---|---|
| 40 principles | 39 problem parameters and Matrix | 2 |
| 76 Standard Solutions | Su-Field Model | 2, 3, 4 |
| Effects | 30 TRIZ standard functions | 2, 3, 4 |
| TE patterns and paths | TE patterns | 2, 3, 4 |

Su-F is a graphical model of a minimal working technique in TRIZ [4]. Figure 1 is a simple description of Su-F Model. Su-F Model provide a fast, simple description of subsystems and their interaction in a technical system via a well-formulated model of the technique in which all subsystems, inputs, and outputs are known or can be quite easily determined. Su-F Model is one of the most important analysis tools in TRIZ, usually used in ARIZ with Standard Solutions.

The use of the Standard Solutions begins with abstracting an initial Su-F model (ISM) from a freely worded description of a problem or situation. Generally, Su-F analysis is adapts to any situation, and figure 2 shows the analysis process [3].

According above, the CAI system based on Standard Solutions should include the Substance-Field Model.

**Fig. 1.** A simple Su-F Model

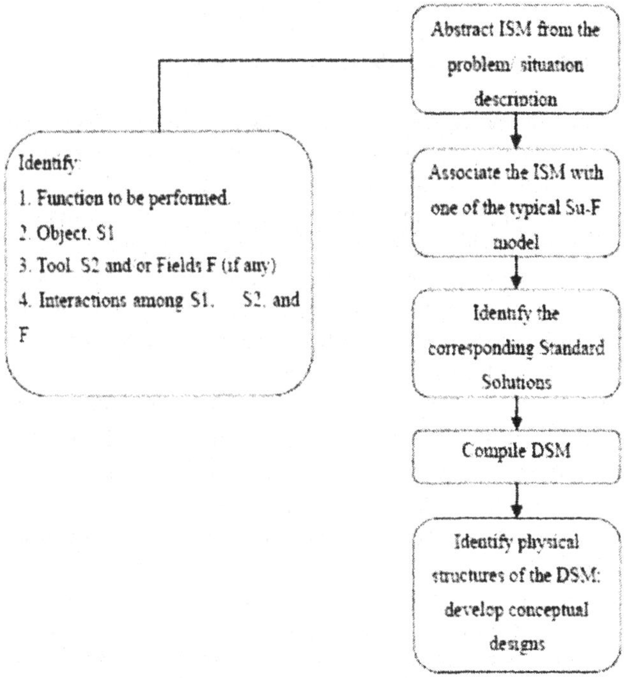


**Fig. 2.** Building Su-F Model using the Standard Solutions [3]

If a technology system, which needs a certain improvement, is modeled as a typical Su-F model, then a corresponding typical transformation can be used to improve the system. Typical Su-F model transformations are called Standard Solutions to solving Problems or, simply, the Standards [3]. Altshuller discovered various engineering problem from different domains can be solved by generic conceptual approaches, which is Standards. They are grouped into 5 or 3 large classes as Table 2.

The 76 Standard Solutions are used to solve relatively common optimization problems. Further more, they are useful for level three inventive problems. Typically, the Standard Solutions are used as a step in ARIZ, after the Su-F Model has been developed and any constraints on the solution have been identified.

Standards and Su-F are linking tightly each other. Before use Standards, a Su-F Model of technical problem must be constructed mostly.

**Table 2.** The Standard Solutions categorization

| | | |
|---|---|---|
| 1. Improving the system with no or little change | | 13 |
| 2. Improving the system by changing the system | System modification | 23 |
| 3. System transitions | | 6 |
| 4. Detection and measurement | System detection and measuring | 17 |
| 5. Strategies for simplification and improvement | Application of Standards | 17 |
| | Total: | 76 |

## 2.2 Algorithms for using the Standard Solutions

Standards are the precepts of synthesis and transformation with the aim to overcome or circumvent technical and physical contradictions [4]. According to figure 2, the Standard Solutions are typical Su-F model transformations. When the ISM's have been built, they must be associated with the available typical Su-F models. Then these Su-F models must be related to one or more of Standard Solutions that "prescribe" transition to more advanced, desirable Su-F models (DSMs). During this process, the appropriate Standard Solutions should be choosing. Some algorithms for choosing the right Standards had been designed.

Joe Miller and Ellen Domb have developed a flowchart for Using the 76 Standard Solutions in the Proceedings of TRIZCON2001. The flowchart organizes use of the 76 Standard solutions into three main pathways: system improvement, solutions for Measurement and Detection, and use of the Standard Solutions for Forecasting opportunities for change [5]. This flowchart is helpful for designer to comprehend the structure of Standard Solutions System. Designer can get a suggestion of class or group level using this flowchart.

Reference 3 thought the Standards are organized in a set built according to the logic of evolution of Su-F models: simple Su-F models →complex Su-F models →intensified Su-F models [3]. So, it provided an algorithm flowchart organized with the logic. This algorithm deal with initial Su-F models in distinguishing detection/measuring problem and non-detection/measuring problem. Based on the transform of Su-F model, user can obtains a right Standard Solution.

Another detailed algorithm was mentioned in reference 4. That algorithm is more detailed and effective than above for choose each item of Standard Solutions. It begin with construct a model of the problem, through a series of opinion according constraints or restrictions, then confirm one or more Standard Solutions for solving problem.

## 2.3 An applicable algorithm for CAI software

All above of the algorithms or methods is useful for designer when they use Standard Solutions to solve innovative problem. When a CAI system based on Standard Solutions was developed, those algorithms must be adjusted to adapt to computer environment. We designed a new algorithm that integrated the advantages of previous and was suitable for computer procedure. Its simplified flowchart shows in figure 3.

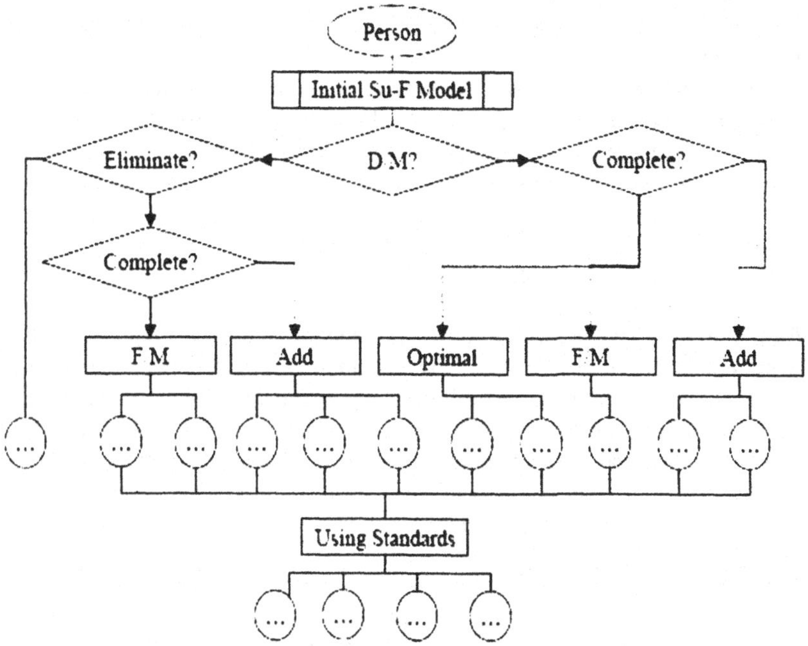


**Fig. 3.** The simplified flowchart of algorithm using in CAI system

This algorithm has some features:

a) Description of constraints is exact;
b) Cover with all the 76 Standard Solutions;
c) Arrangement of group or class or item is clear;

# 3 The CAI System Based on Standard Solutions

## 3.1 Situation of CAIs Based on TRIZ

Computer Aided Innovation software based on innovational theory is a kind of CAD that is suitable for conceptual design stage [6]. For TRIZ has systematized configuration, multi tools to analyze and model problem, mighty knowledge database to provide example or analogy cases, TRIZ-based CAI software have been developed rapidly and widely than other innovational theory.

In 1991, a TRIZ-based software package had been developed by the Invention Machine Corporation. TRISolver 2.1™ is the early English-language version of the

CAI software from TRISolver Consulting Group in Germany. The Invention Machine in America had issued TechOptimizer, and than upgrade to Goldfire Innovator. Another America firm Ideation International developed the Innovation WorkBench (IWB) and so on. Those English-language software of CAI integrate the concepts, principles and tools of TRIZ. Those CAI software were applied in practice and made great success[7]. In China, InventionTool had been commercially launched by the 3rd version.

Some specialized functional CAI system were developed depend upon TRIZ tools, such as inventive principles, technology evolution, effect, and technology maturation forecast and so on. Usage of them can make designer agile, convenient and quick solve some little problem. Those systems have growing potentiality.

### 3.2 The Structure of Standard Solutions CAI System

Basing on the above algorithm for using Standard Solutions, a Standard Solutions CAI system is developed. In the CAI system it is realized not only to support technical knowledge but to solve problem using in Standard Solutions systemically.

Standard Solutions CAI system is a module of InventionTool CAI software, also can be used individual. TRIZ is innovational theory based on knowledge, and same to the Standard Solutions as a part of TRIZ. The whole system should include four components: Substance-Field, Standard Solutions, Algorithm and database, as shown in figure 4.

Before using the Standard Solutions to solve problem, designers should be expert in modeling by Su-F and in all the Standards. In the system, Substance-Field component show the definition and graphic representation for designer. Standard Solutions component enumerate items one by one with examples. Algorithm provides the process solving problem using Standards. Database stores the examples which can be analogized, and it can be add new examples.

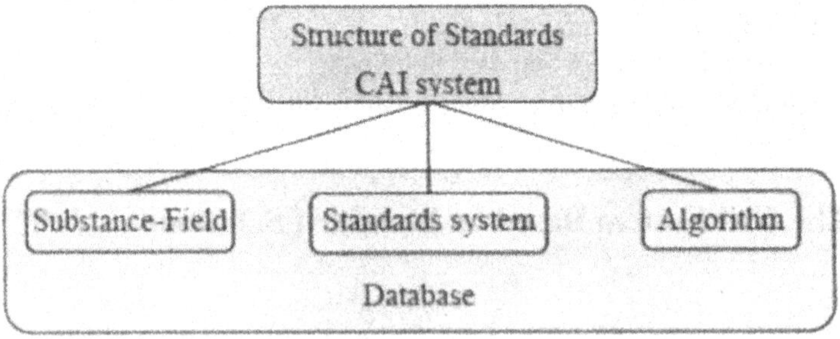


**Fig. 4.** Structure of Standard Solutions CAI System

### 3.3 Method and Tools Used in Software Development

In term of the lifecycle of software, a software development process is fallen into 6 steps: analysis, design, implementation, module test, system test and executing. The Standard Solution system follows the same process. Object-Oriented method [8] is the mainstream method to develop software. UML method integrated and expended the advanced characteristics of some other Object-Oriented methods. The system analysis obeys UML, and implement with VC++ language and XML.

The Unified Modeling Language (UML) is a modeling language for specifying, visualizing, constructing, and documenting the artifacts of a system-intensive process. [9] Before to program codes, the models of software system can help software designer study the various constructions and schemes conveniently. Through construct a series of models or diagrams, "what" is required of a system, and "how" a system may be realized was shows before software designers. .UML is applicable in different stages of system development, from description of required specification to system test and maintenance.

After analyzing the Standard Solutions CAI system, software was coded with VC++ language. The interfacial view of this software, which is friendly for users, was created by VC++ combines with XML+XSL. For the CAI system has a large database, XSL is a suitable tool to deal with data transfer and data store. XML (eXtensible Markup Language) is used to develop the interface of the software.

## 4 UML Models in the Standards CAI System

During the development of CAI software product, static models such as Use Case Diagram, Class Diagram, Object Diagram, Package, Component Diagram and Deployment Diagram built based on UML were used to construct the configuration of system. For the components of system, dynamic models such as State Diagram, Sequence Diagram, Collaboration Diagram and Activity Diagram were used to describe the behaviors of system components.

Before develop the Standard Solutions CAI system, Use Case model should be construct, like shows in figure 5. It shows what is actor or user recognized of system and how to operate it.

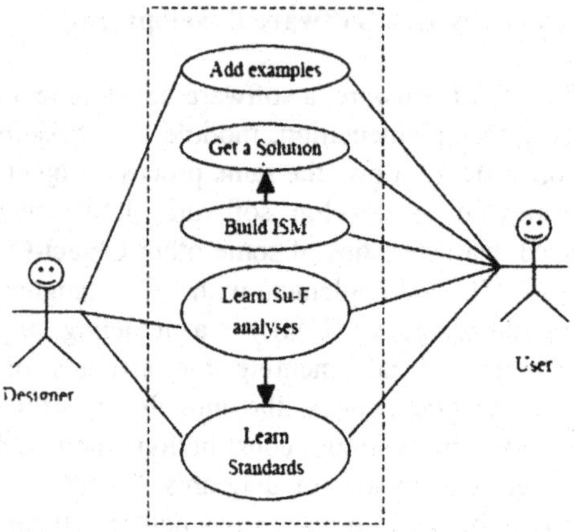


**Fig. 5.** The Use Case Diagram of Standards CAI system

During the software was programmed, many kinds of UML model had been built. They show different relationships and operation. Figure 6 is a Activity model which shows add new examples to system. The activity of add was shown clearly.

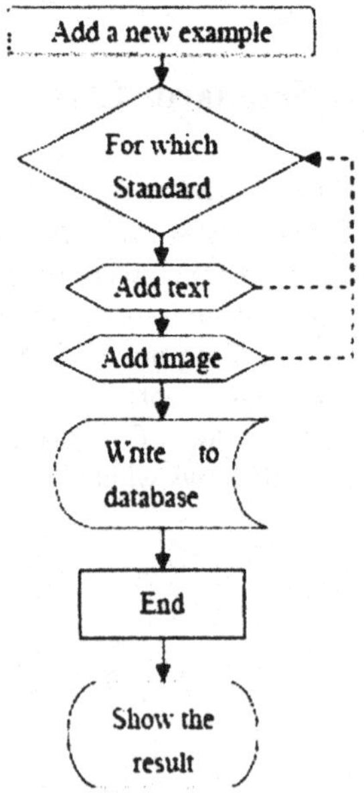


**Fig. 6.** The Activity Diagram of add new example to system

All above of the UML model is helpful to develop software, improve the efficiency of programming.

# 5 Conclusions

Standard Solutions for innovation firstly models a technical or process problem by Substance-Field (Su-F) Analysis, then synthesizes and converts the problem to acquire a solution. A CAI system based on Standard Solutions was developed that include Su-F model, Standards and theirs usage. The algorithm for using Standard Solutions was improved. The development supported by UML method, VC++ language and XML. Convenient usage and friendly interface have great help to innovation design. The method for development of CAI product is instructive exploration which had been proved to be efficient and feasible during the process of practice.

# 6 Acknowledgement

The research is supported in part by the Chinese Natural Science Foundation under Grant Numbers 50675059 and Chinese national 863 planning project under Grant Number 2006AA04Z109.

# 7 References

1. Altshuller G., The Innovation Algorithm, TRIZ, systematic innovation and technical vreativity, Technical Innovation Center, INC, Worcester, 1999
2. John Terninko, Ellen Domb, Joe Miller, The Seventy-six Standard Solutions with Examples, The TRIZ journal, February, 2000, http://www.triz-journal.com.
3. Victor Fey and Eugene Rivin, Innovation on Demand, New Product Development Using TRIZ, Cambridge University Press, 2005
4. Savransky S D, Engineering of Creativity, New York: CRC Press, 2000
5. Joe Miller, Ellen Domb, Ellen MacGran, John Terninko, Using the 76 Standard Solutions: A case study for improving the world food supply, The TRIZ journal, April, 2001, http://www.triz-journal.com.
6. Tan Runhua, Progress of some problems in product design for innovation, Chinese Journal of Mechanical Engineering, Sep., 2003
7. Yang Bojun, Tan Runhua, Tian Yumei, Development of a CAI system of Standard solutions based on TRIZ, Proceedings of PROLAMAT2006, IFIP TC5 International Conference, Shanghai, 2006
8. Hassan Gomaa, An Object-Oriented domain ananlysis and modeling method for software reuse, 0073-1129-1/92, 1992 IEEE
9. Grady Booch, James Rumbaugh, Ivar Jacobson. The Unified Modeling Language User Guide. Addison- Wesley, Reading, Mass, 1999

# Computer-Aided Patent Analysis: finding invention peculiarities

Gaetano Cascini[1], Davide Russo[1], Manuel Zini[2]
1 University of Florence - Department of Mechanics and Industrial Technologies, Methods and Tools for Innovation Lab, Florence, Italy
{name.surname}@unifi.it
2 DrWolf srl, Florence, Italy
mlzini@drwolf.it

**Abstract**. The application of standard Information Extraction techniques to Patent Analysis has several limitations partially due to the difference existing between patents and web pages, which are the object of the biggest majority of information search. Indeed, while in other fields customized processing techniques have been developed, the number of studies fully dedicated to patent text mining is very limited and the tools available on the market still require a relevant human workload. This paper presents an algorithm to identify the peculiarities of an invention through an automatic functional analysis of the patent text; as a result a ranked list of components and functions is provided as well as a selection of meaningful paragraphs disclosing the details of the invention. An example related to laser irradiation devices for medical treatment clarifies its basic steps.

## 1 Introduction

Today's text mining research activities are mostly dedicated to web content mining and encompass resource discovery from the Web, document categorization and information extraction from Web pages. The latter aims at the identification of the most relevant portion of a document and typically is based on the analysis of anchortexts, i.e. the visible part of hyperlinks. The rationale is that the larger is the number of anchortext terms in a sentence, the more relevant the sentence is likely to be, since it is supposed that the relevant sentences in the destination page are related with the anchortext in the source page [1]. Nevertheless, it is clear that this kind of approach is not applicable to patent analysis: the typical link between these documents is the citation, in which the anchortext is just the patent reference number, thus it is not related to conceptual details of an invention.

Another typical strategy consists in building ontologies to map terms relationships in a specific field, as performed in [2]. Such an approach is highly time consuming and thus it is still not widely applied.

---

*Please use the following format when citing this chapter:*

Cascini, G., Russo, D., Zini, M., 2007, in IFIP International Federation for Information Processing, Volume 250, Trends in Computer Aided Innovation, ed. León-Rovira, N., (Boston: Springer), pp. 167-178.

Moreover patent examiners are skeptical about the adoption of software instruments substituting traditional Boolean search engines and manual efforts to perform prior art analyses. Nevertheless, they consider a top level priority the development of means to reduce the number of document to browse [3]. It can be stated that with the same purpose of reducing human involvement, also the reduction of the amount of text to be read from each document still is an essential goal.

In the past the authors have developed algorithms and tools for patent analysis aimed at:

- translating the description of an invention into a conceptual functional map [4];
- identifying knowledge flows between different fields of application [5];
- investigating the properties of Small World Networks as a base for Computer-Based idea generation system [6].

Among the crucial issues that emerged during those studies, the most challenging one is the identification of the most relevant part of a patent, i.e. the paragraphs disclosing the invention peculiarities. In other words, a relevant research goal is the capability to identify selected excerpts of the patent description, connected each other through the functional map of the invention, so that an expert in the field can focus his/her attention just on few sentences, instead of reading the whole text.

In [7] regular expressions and the analysis of the detail level of the description were presented as a means to achieve such a goal.

In this paper the adoption of a *tf-idf* (term frequency-inverse document frequency) ranking approach, to be performed after building a specific Thesaurus, is proposed as a means to highlight the relevant details of an invention and their disclosing sentences. The second section of the paper summarizes the previous results obtained by the authors and a comparison with other text-mining approaches is reported. Then, the proposed algorithm is detailed in section 3 and an exemplary application related to tumor ablation devices is shown in section 4 to demonstrate the efficiency of the proposed approach. A final discussion and opportunities for further developments conclude the last chapter.

## 2 Patent mining, related art

Patent mining is the branch of text mining technologies dedicated to the extraction of relevant information from patents and to their categorization. Indeed, just a few specialized tools fully dedicated to patent analysis exist, while typically general purpose text mining applications are adopted in combination with traditional Boolean search engines.

Commercially available patent databases provide basic means for information retrieval and citations tracking, but patents searches are still time consuming and require big efforts to be accomplished. In facts, citation analyses are the most used techniques for identifying within a company's patent portfolio the small number of valuable, high-impact patents against the large number of patents of marginal importance [8]. It is believed that a statistical analysis of the rate of publication of patents pertaining to a certain field or assigned to a certain company, provides information about technology maturity and corporate technology strategies. Typically, the analysis is performed by counting in an online database the number of patents issued annually in a set of calendar years [9]. Besides, it normally takes five or more years from publication before a patent begins to be cited to any great extent. In general, 70% of all patents are either never cited, or cited only once or twice, so that even five citations place a patent in the top few percent of cited patents [10].

Therefore, the analysis of the free textual description is assuming a greater relevance for getting major advantages from disclosed inventions.

Text Mining applications provide effective means for content searches in the textual fields of electronic documents databases, but also the most recent works are not tailored for patent analyses as [11] and too often require a deep expertise about how to gain major advantages from this technology. Some special features are available in the Invention Machine Goldfire platform [12], mainly related to the application of syntactic parsing capabilities: each sentence is translated into a SAO triad (Subject, Action, Object), in order to produce a classification of the concepts contained in a patent description. Nevertheless, as well as for more traditional keywords based tools, no systems are available on the market for capturing the role of a component in an invention and for grouping patents according to their peculiar functionalities apart from their fields of application.

More specifically the following features aimed at speeding-up patent analysts activity still lack on the market:

- identifying the architecture of the claimed invention, distinguishing the functional (semantic) role of each component;
- identifying invention peculiarities as a means for providing an automatic extraction of the core of the patent; (it is worth to mention that too often the patent abstract is very low informative);
- clustering technical solutions according to the way a function is accomplished apart from the field of application (therefore providing proper means for technology transfer);
- allowing easy and effective queries by means of a multi-language taxonomic knowledge base so that search results do not depend on patent language and/or the use of synonyms, hyperonyms, meronymes etc.

Among the other activities, the authors have addressed the first issue by developing a novel analysis approach, disclosed in [4]. The proposed algorithm is capable of performing the functional analysis of an invention automatically, by processing the description and the claims of the related patent.

The algorithm basically consists in: (i) identifying the components of the invention; (ii) classifying the identified components in terms of detail/abstraction level and their compositional relationships in terms of supersystem/subsystem links; (iii) identifying positional and functional interactions between the components both internal and external to the system.

The components identification is performed taking into account that all the components must be referenced univocally to be identified in the patent figures. The following step of the analysis process is dedicated to the search of descriptive locutions (i.e. sentences containing verbs like “to form”, “to constitute” etc.) and specification expressions (like “the gripper of the pivot arm”) in order to identify subsystem/supersystem relationships, hence defining a hierarchy of detail/abstraction levels. Finally, positional and functional interactions between the identified components are determined by filtering, from the list of SAOs provided by a syntactic parser, the triads containing irrelevant verbs (i.e. verbs like "to refer", "to show", as well as any other verb not describing some function or action).

Fig. 1 shows an exemplary application of the algorithm: the list of components and their hierarchical structure is represented by a tree; their functional interaction are mapped through a directed graph. These outputs, obtained fully automatically, i.e. without any interaction with the user, can be further processed in order to identify the peculiarities of the invention, as described below.

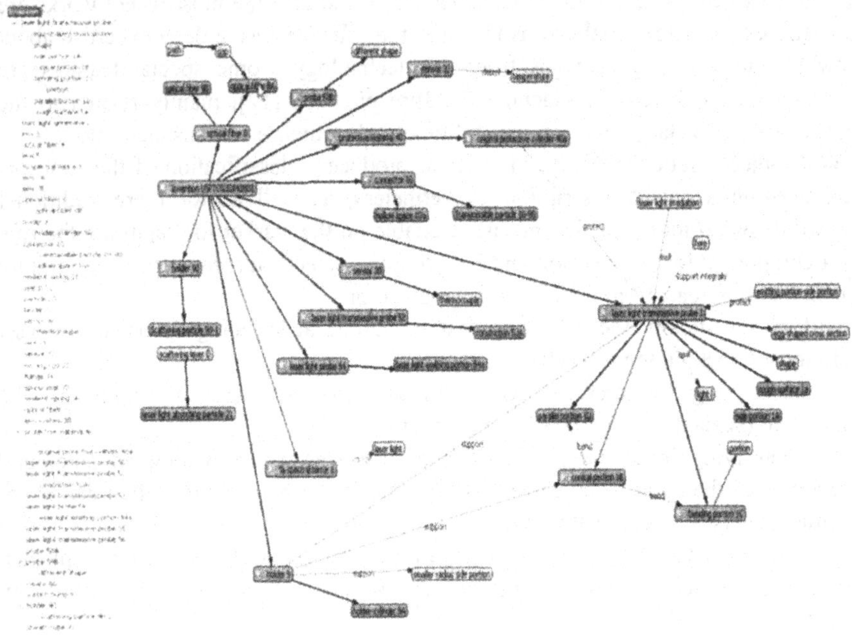

**Fig. 1.** Functional analysis of US patent 5,328,488 "Laser light irradiation apparatus for medical treatment": list and hierarchy of components (left), functional map (right).

## 3 The functional subtraction algorithm

The rationale of the algorithm proposed in this paper is identifying the peculiarities of an invention by isolating the components and/or the functions of the system that differ from the state of the art. It is clear that the comparison among different patents should be independent from the language style of the inventor.

In order to fulfill this goal, the authors have developed an algorithm for building a Thesaurus of a specific set of patents.

It is worth to mention that during the components identification phase, alternative denominations can be found for each element of the system if several multi-words are referred to the same component reference number. For example in the patent US 6,161,390 – "Ice maker assembly in refrigerator and method for controlling the same", the component 52 is called by the writer of the patent both "ice tray" and "ice container".

If two or more patents have a component sharing a common denomination, it is assumed that all their alternative denominations can be considered as synonyms forming a single Thesaurus entry. With this assumption the Thesaurus can be automatically built.

Such an assumption can be considered valid as far as the selected patents belong to a specific field of application, i.e. the same IPC class. Since the proposed algorithm for Thesaurus construction doesn't require any contribution of the user, it is clear that noisy terms may appear in the synonyms lists; nevertheless, according to the authors' experience, these noisy terms don't compromise the overall quality of the analysis. However, it is obvious that a manual validation of the extracted synonyms lists can contribute to a further improvement of the Thesaurus reliability.

Thus, when analyzing a set of patents related to a specific product/process, it is proposed to compare their components and functional interactions automatically extracted by means of the technique summarized in the previous section, taking into account the synonym lists gathered in the Thesaurus. The relevance of a component or a function $X$ in a patent $k$ is estimated according to the following formula, derived from the well known *tf-idf* weighting criterion [13]:

$$Score(X\ in\ Patent\ k) = \frac{\#of\ occurrences\ of\ X\ in\ k}{\max \#of\ occurrences\ in\ k} \log\left(\frac{\#of\ patents\ in\ the\ set}{\#of\ patents\ containing\ X}\right) \quad (1)$$

As stated above, the count of the number of occurrences and the count of the patents containing a certain component/function is performed assuming the equivalence of the synonyms gathered in the Thesaurus.

More in details, the proposed algorithm consists in the following steps:

1. Gather a set of patents related to a specific product/process;
2. Perform an automatic functional analysis of each patent according to the algorithm mentioned in section 2;
3. Build the Thesaurus of the patent set by identifying correspondences among the alternative denominations of the components of each patent;
4. Evaluate the score of each component of each patent according to the formula (1); evaluate the score of each functional interaction, i.e. each triad component-verb-component, according to the formula (1);
5. Identify the components at the highest detail level, i.e. the deepest leafs of the hierarchical tree built at step 2;
6. Extract from each patent the excerpt containing the top ranked components/functions.

## 4 Exemplary application of the proposed algorithm

A brief case study is here reported in order to clarify the algorithm described in the previous section and to demonstrate its validity. Instead of proposing the analysis of a big number of inventions that makes the manual validation rather critical, the selected example is related to a small set of selected patents: the test set is here constituted by six patents related to laser irradiation devices for medical treatment. The set has been chosen because these patents are very similar to each other, they share the same title and the same inventor and, most of all, they deal with small improvements of the same device, thus the identification of their peculiarity requires a careful analysis also by a reader skilled in the art.

The test set is constituted by six patents titled "Laser light irradiation apparatus [for medical treatment]", assignee S.L.T. (Surgical Laser Technologies), issued from March 2003 to March 2005 and belonging to the US Classes 606/16 and 606/17: US 5193526, US 5209748, US 5290280, US 5328488, US 5496307.

**Table 1.** Patent US 5,328,488 "Laser light irradiation apparatus for medical treatment": excerpt from the list of components and their alternative denominations. Due to space limitations just the components with multiple denominations have been kept.

| Component – ID Number | Alternative denominations |
|---|---|
| Laser light transmissive probe 1 | laser light transmissive probe; probe; right side laser light transmissive probe; opposite laser light transmissive probe; laser light penetrating probe; transmissive probe; light transmissive probe; penetrating probe |
| rough surface 1a | rough surface; notch |
| optical fiber 8 | optical fiber; single optical fiber |
| holder 9 | holder; pinching holder |
| holder-cylinder 9A | holder-cylinder; holder |
| particle 20 | particle; laser light scattering particle; scattering particle |
| wire 31 | wire; lead wire |
| spring seat 35 | rod; spring seat |
| resilient spring 36 | resilient spring; spring |
| protective material 40 | metal protective material; protective material; material |
| original protective cylinder 40a | original protective cylinder; cylinder |
| laser light transmissive probe 53 | laser light transmissive probe; probe |
| laser light probe 54 | laser light probe; penetrating probe; probe |
| laser light emitting portion 54a | laser light emitting portion; flat emitting portion |

The application of the functional analysis algorithm (step 2) leads to the extraction from each patent of the followings:

- List of components and their alternative denominations (Table 1);
- Hierarchical tree of the components (Fig. 1, left);
- Functional interactions graph (Fig. 1, right).

Due to space limitations it is not possible to report the outputs arising from the analysis of each patent; besides, when dealing with a high number of patents, it is not convenient to start reading the details extracted from the individual patents, but it is suggested to start with an overall survey according to the ranking determined through the steps 4, 5. The bottom ranked components, i.e. those shared by the majority of the patents under comparison, constitute the common core of the examined system. The Table 2 reports the common components of the test set and their alternative denominations provided by the Thesaurus. It can be stated that the novelty of the selected patents doesn't reside in the introduction of those components, while it might happen that the novelty consists in a modification of one of them aimed at providing a special property/functionality. The latter case can be identified as explained below.

In order to understand the rationale of the *tf-idf* criterion here applied, it is useful to focus the attention on the second and third column of Table 3, representing the components with the highest idf score and the exclusive forms/multi-words extracted respectively. It can be verified that most of the top idf-scored components are closely linked to the novelty disclosed in those patents. In facts, the "ballon" in US5,193,526, the "nipple" in US5,209,748, the "clads" in US5,290,280, the "constrictor" in US5,328,488 and the "fluid outlet" in US5,496,307 constitute the invention peculiarity or, at least, are strictly related to it. Those components point directly to the core of the invention and a person skilled in the art about laser irradiation devices for medical treatments will immediately understand what the patent deals with.

**Table 2.** Common components shared by the majority of the patents of the test set.

| **Reference Component** | **Alternative denominations** |
|---|---|
| Laser light | laser light generator, right side laser light, constriction, portion, plural optical fibers, expose core |
| Fiber optic | optical fibers, original optical fibers, single optical fiber, original optical fiber, core |
| Layer | gold plate layer, laser light reflective layer, reflective layer, reflection layer, surface layer, concave surface |
| Probe | transmissive probe, penetrate probe, rough surface, emitter, laser light emitter, cylindrical-shaped emitter |
| Thermocouple | lead wire |
| Wire | guide wire |
| Tube | flexible protection tube, protection tube, holder tube, synthetic resin holder tube, main tube, core ,support tube, conductive tube, hole |
| Sheath | flexible sheath, sheath tube |
| Holder | metal holder, sleeve-like connector, hollow space |
| Lens | impinge lens, lens system |

**Table 3.** Extraction of invention peculiarities through combined criteria.

| US patent | idf score | Exclusive forms, multiwords | Detail level |
|---|---|---|---|
| **5193526** | Balloon 11 | Hole 2A<br>Impinge face 1A | |
| **5209748** | Nipple 3 | Body 6A<br>Screw hole 6C | |
| **5290280** | Emitter 20<br>Clads 1A,B<br>Handle 5<br>Clad material 10 | Cylindrical-shaped emitter 20A<br>Knife-shaped flat emitter 20B<br>Hook-shaped flat emitter 20C<br>Claw-shaped emitters 20D<br>Sickle-shaped emitter 20E<br>Grip handle 5C<br>Impinge lens 3 | Emitter 20 |
| **5328488** | Binder 22<br>Pump 61<br>Switch 12<br>Constrictor 52 | Pinching holder | Laser light penetrating probe 58A |
| **5496307** | Covering 24<br>Fluid outlet 24a<br>Fastener 20 | Metallic fastener | Optical fiber 8B |

The exclusive forms/multi-words may draw the attention to properties and characteristics of invention details; for example, from US5,290,280 a number of characteristic shapes of the emitter are highlighted: cylindrical, flat knife, flat hook, claw, sickle. Further relevant features can be extracted by identifying the components at the highest detail level (step 5), i.e. the deepest leafs of the hierarchical tree built at step 2. The assumption here is that the description involves specific sub-components of the invention only if they are meaningful to the explanation of the invention itself.

The whole set of selected components and functions resulting from the steps 4 and 5 of the proposed algorithm can be used as seeds for a Content Analysis [14] of each patent The output is a selection of paragraphs where the top-ranked concepts are more represented (step 6). Again it can be stated that an expert in the field will be able to understand the content of those paragraphs without reading the whole document, at least for recognizing the relevance of the patent and its core novelty.

The top ranked paragraphs of the test set here adopted are reported in Table 4 and compared with the corresponding abstracts in order to demonstrate their higher informative content.

**Table 4.** Selected paragraphs of the test set.

| **US 5,193,526** 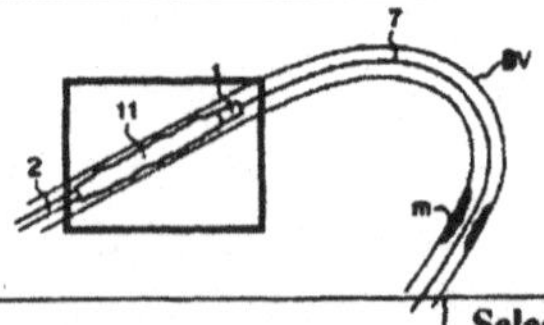 | 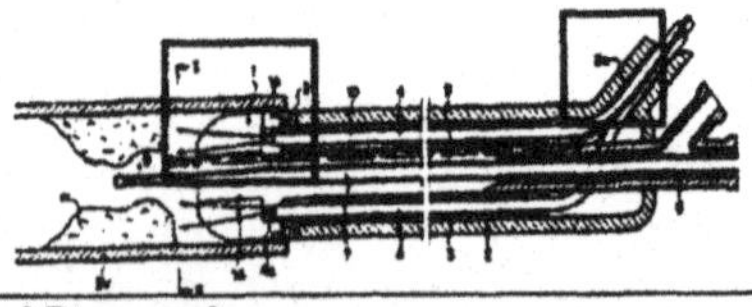 |
|---|---|
| **Patent Abstract** | **Selected Paragraphs** |
| A laser light irradiation apparatus used for medical treatment of living tissues. According to a preferred embodiment, the apparatus comprises a probe and a plural number of optical fibers. The optical fibers surround the axis of the probe. Laser light goes through each optical fiber and is applied to the probe. Then, the laser light is emitted from the probe to | • By the laser light irradiation, the stricture part m is burnt off to widen the inside of a blood vessel. If desired, as shown in FIG. 4, pressurized air or pressurized liquid is sent into a **balloon 11** connected between a probe 1 and a main tube 2, thus, the **balloon 11** is expanded and press the stricture part m. As a result, together with the above mentioned burning off the inside of the blood vessel by the laser light irradiation, the |

| | |
|---|---|
| uniformly irradiate the tissues, and if desired, against the tissues over a broad area. Further, a guide wire and/or a lead wire for detecting a temperature can be placed so as to be coaxial with the probe. Therefore, a perforation of a normal part of the blood vessel can be prevented. | stricture part m can be broken mechanically.<br>• Each tip portion of the optical fiber 1 is exposed to a core 4a. Each core 4a is adjacent to the back end face or the **impinging face 1a** of the probe 1.<br>• Each optical fiber 4 is inserted into the main tube 2 from an inserting **hole 2a**. |

## US 5,209,748

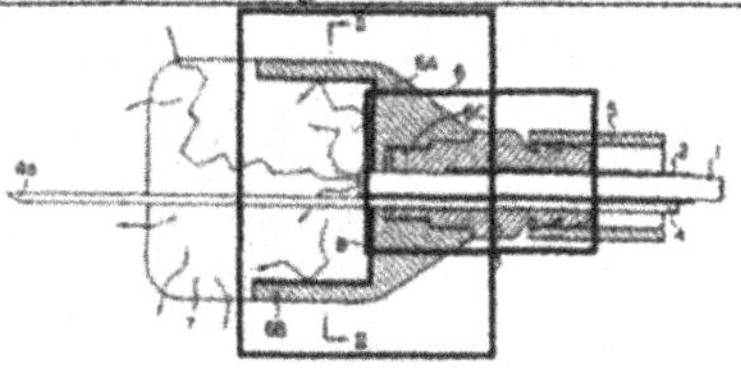

| Patent Abstract | Selected Paragraphs |
|---|---|
| A laser light irradiation apparatus used for medical treatment of tissues. According to a preferred embodiment, the apparatus comprises a probe, an optical fiber feeding laser light into the probe and a lead wire for detecting a temperature being inserted through and projecting from the probe. Then, the probe contains laser light scattering particles for uniform irradiation of the laser light against the tissues. Further, the probe is fabricated from a laser light tramissive synthetic material, and the fore end of a core of the optical fiber and the inserting part of the lead wire are in the synthetic material of the probe for easy molding for this apparatus. | • The fore end portion of the optical fiber 1 is inserted through a **nipple 3**, which is fabricated from a synthetic material such as polyethylene and the like. A lead wire 4 detecting a temperature having a thermocouple 4a at its fore end is provided alongside the optical fiber 1 and is also inserted through the **nipple 3**.<br>• The holder 6 comprises a **body 6A**, which is tapered toward its back end, and a sleeve-like connector 6B, which has a hollow shape and which is projected from the **body 6A**. The screw of the **nipple 3** is adapted to mate with a connecting screw hole 6C of the holder 6 for connection. The optical fiber 1 and the lead wire 4 for detecting the temperature are inserted through the **body 6A**. |

## US 5,290,280

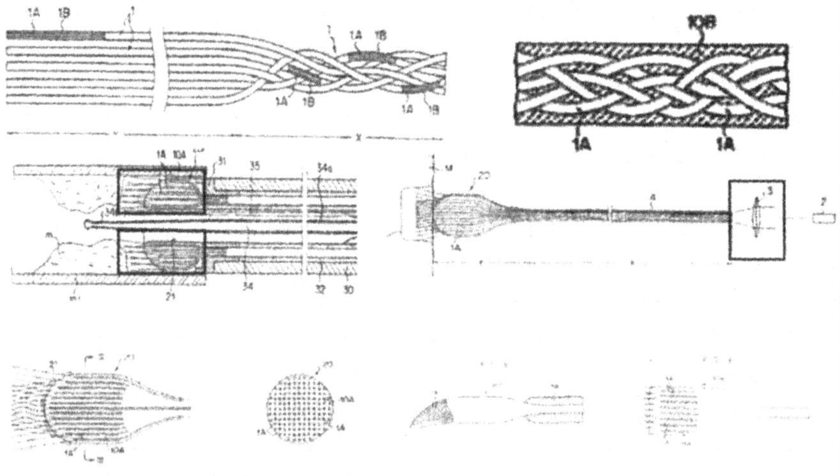

| Patent Abstract | Selected Paragraphs |
|---|---|
| A laser light irradiation apparatus for medical treatment of living tissues, a preferred embodiment, comprises a laser light emitter and plurality of optical fibers. The fore end portion of each optical fiber is exposed to form an exposed light emitting core. The exposed cores are surrounded by a clad-material serving as the laser light emitter in order to reduce power loss of the laser light. Also, since there is no space between the emitting face of the optical fiber and the impinging face of the emitter, a cooling fluid is not required to pass through. The laser light is emitted from the emitter to | • Each original optical fiber 1 has a core 1A and a **clad 1B** surrounding the core 1A. Then, while at the base portion X, the fibers 1 are twisted in an irregular manner, the twisted original optical fibers 1 at the base portion X are heated to a temperature which is substantially the same temperature as or higher temperature than the melting point of the **clad 1B** and which is lower temperature than the melting point of the core 1A. Then, at the base portion X, the **clads 1B** of original optical fibers 1 are moulded to be one **clad 10B**, which includes the twisted cores 1A....<br>• As a result, a laser light **emitter 20** composing the resulting **clad-material 10A** and the number of cores 1A, which are arranged in parallel and which are included in the **clad-material 10A**. The shape of the laser light **emitter 20** corresponds to the shape of a container including the **clad-material 10A**. For example, as shown in FIG. 1, if the container has a constriction at the back end of the **emitter 20**, the shape of the laser light |

| | |
|---|---|
| irradiate uniformly against the tissues, and if desired, against the tissues having a broad area. Further, a guide wire and a lead wire detecting a temperature can extend coaxially through the emitter. Therefore, a perforation of a normal part of the blood vessel can be prevented. To provide a more uniform power level distribution of the laser light, the optical fibers at the base portions are twisted. | **emitter 20** should be provided with an open having an inner diameter corresponding to the diameter of the constriction.<br>• A **grip handle 5C** is provided at the back portion of the **emitter 20D** and can be operated with a restoring force.<br>• The laser light irradiation apparatus of this type described above is used as follows. First, laser light fed from a laser light generator 2 goes through an **impinging lense 3**.<br>• In the present invention, the **emitter** having several kinds of shapes can be applied. There are, for example, a **cylindrical-shaped emitter 20A** having a flat emitting face as shown in FIG. 6, a **knife-shaped flat emitter 20B** as shown in FIGS. 7 and 8, a **hook-shaped flat emitter 20C** as shown in FIG. 9, **claw-shaped emitters** 20D as shown in FIGS. 14, 15, 16 and 17, a **sickle-shaped emitter 20E** as shown in FIGS. 18, 19, 20 and 21. |

## US 5,496,307

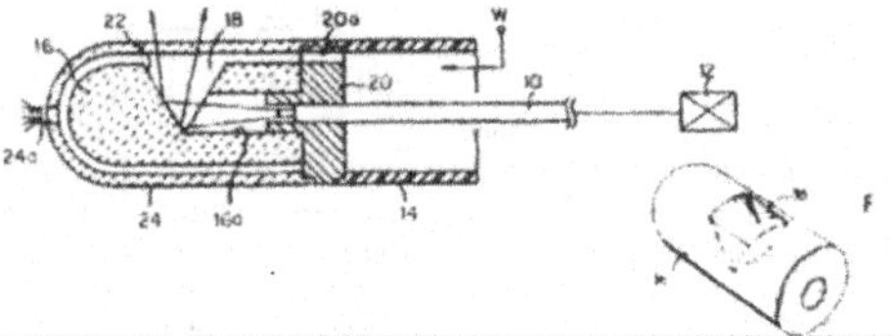

| Patent Abstract | Selected Paragraphs |
|---|---|
| A laser light irradiation apparatus for medical treatment by irradiating an object tissue with laser lights transmitted through an optical fiber or fibers comprises a laser light reflector provided in front of the laser light emitting end of said optical fiber for reflecting the laser lights in a lateral direction of the apparatus, a covering which covers the reflector and is capable of transmitting the laser lights at least at the side portion thereof, wherein fluid is continuously supplied to a space between the covering and said reflector. | • A protection tube, separate from the covering, surrounding said optical fiber, said protection tube being linked with said **covering** by means of a **fastener** having a through hole, said **covering** being formed with a **fluid outlet** through which the supplied fluid having passed through said through hole and said protection tube can be discharged.<br>• A **covering 24** which is made of, for example, light transparent ceramics such as heat resistant glass is linked with the front end of the protection tube 14 via the metallic **fastener 20** to enclose the reflector 16 therein. The **covering 24** is formed with a **fluid outlet 24a** at the front end thereof. Fluid, such as cooling water is supplied into a space between the protection tube 14 and the **optical fiber 10**. The flange of the **fastener 20** is formed with one or more through-holes 20a. |

## US 5328488

Figure 17

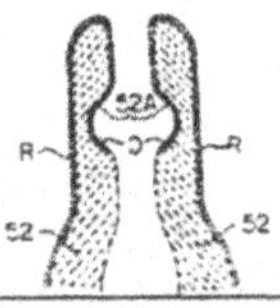

Figure 18

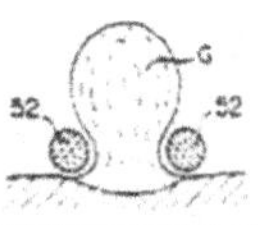

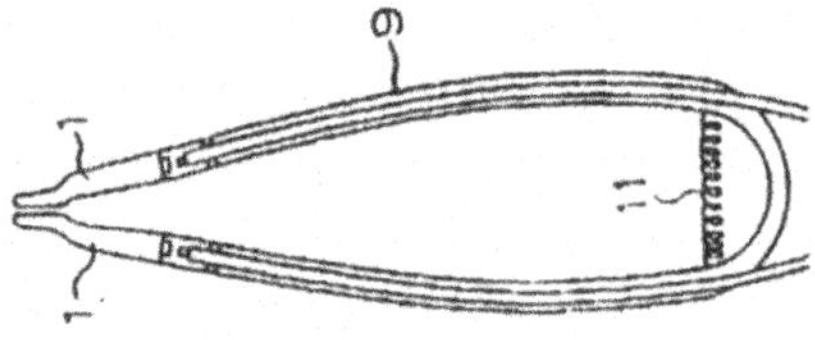

| Patent Abstract | Selected Paragraphs |
|---|---|
| Laser light apparatus for medical treatment to permit amputations, incisions, vaporization of living tissues of an animal such as a human body, thermal therapy and the like. This apparatus consists of a laser light generator, a laser light transmissive probe system and a laser light transmitting system. The laser light transmissive probe system is provided with an opposed pair of laser light transmissive probes. The opposed pair of probes can be controlled by a medical operator so as to be moved into or out of contact with each other at their laser light emitting portions. Laser light is transmitted to the opposed pair of probes from the laser light generator through the laser light transmitting system. Then, a target area of living tissues is pinched by the opposed pair of laser light transmissive probes so as to be disposed between the opposed pair of laser light emitting portions. | • Further, the fore end portions of tho optical fibers 8A, 8B are provided in a U-shaped holder, in this embodiment a **pinching holder 9**. The **pinching holder 9** is made of metal and moves like a pincette. The above mentioned opposed pair of laser light transmissive probes 1, 1 are supported integrally by the fore end portions of the **pinching holder 9**.<br>• On the other hand, as shown in FIGS. 17 and 18, when a projected tumor G formed on the surface of the tissues are excised, an opposed pair of laser light transmissive probes 52, 52 provided with pair of **constrictions 52A**, 52A can be used effectively. The opposed pair of laser light scattering layers D, D are preferably formed on the inner surface of the opposed pair of **constrictions 52A**, 52A respectively. |

## 5 Conclusions and further developments

This paper presents an algorithm for patent analysis aimed at the identification of the invention peculiarities. Compared with standard Information Extraction techniques, the proposed approach is strongly based on typical patent features and first of all on the way the components of the invention are referred within the description. Such a characteristic allows to build a Thesaurus of the patent set under analysis that can be used to compare the inventions with reduced dependence from the language style of the author. Moreover, the comparison is not based on keywords directly extracted from the patent description, but relies on the identification of the invention components and their functional interactions.

The output of the proposed algorithm is a list of highlighted components and a small number of paragraphs representing the most relevant portion of the description; this excerpt of the patent is sufficient for an expert in the field to realize the scope of the invention and its core content. The algorithm has been applied to a small set of patents related to laser emitting devices for medical treatments to clarify the process and to demonstrate the advantages of the proposed approach.

The selected paragraphs can be used to improve the efficiency of clustering tools, both reducing the amount of text to be processed and removing the noisy part of the text, as proposed in [15].

A further improvement of the technique can be obtained by organizing the outputs according to a more comprehensive conceptual model, as proposed in [16]. Nevertheless, it is worth to note that in such a paper the authors identify the model elements by means of language patterns strongly dependent on the style of the writer.

## References


1. Chen L., Chue W. L.: "Using Web structure and summarization techniques for Web content mining", Information Processing and Management, vol. 41, 2005, pp. 1225–1242.
2. Krauthammera M., Nenadic G.: "Term identification in the biomedical literature", Journal of Biomedical Informatics, Volume 37, Issue 6 , December 2004, Pages 512-526.
3. Krier M., Zaccà F.: "Automatic categorization applications at the European patent office", World Patent Information, Volume 24, Issue 3 , September 2002, Pages 187-196.
4. Cascini G.: "System and Method for performing functional analyses making use of a plurality of inputs", Patent Application 02425149.8, European Patent Office, 14.3.2002, International Publication Number WO 03/077154 A2 (18 September 2003).
5. Cascini G., Neri F., "Natural Language Processing for patents analysis and classification", Proceedings of the TRIZ Future 4th World Conference, Florence, 3-5 November 2004, published by Firenze University Press, ISBN 88-8453-221-3.
6. Cascini G., Agili A., Zini M.: "Building a patents small-world network as a tool for Computer-Aided Innovation", Proceedings of the 1st IFIP Working Conference on Computer Aided Innovation, Ulm Germany, November 14-15, 2005.
7. Cascini G., Russo D., "Computer-Aided analysis of patents and search for TRIZ contradictions", International Journal of Product Development, Special Issue: Creativity and Innovation Employing TRIZ, Vol. 4, Nos. 1/2, 2007.
8. Breitzman A.F., Mogee M. E.: "The many applications of patent analysis". Journal of Information Science, 28 (3), pp. 187–205, 2002.

9. Bigwood M.P.: "Patent Trend Analysis: Incorporate Current Year Data". World Patent Information, Vol. 19, No. 4, pp. 243-249, 1997.
10. Karki M.M.S.: "Patent Citation Analysis: A Policy Analysis Tool". World Patent Information, Vol. 19, No. 4, pp. 269-212, 1991.
11. Trappey A., Hsua F. C., Trappey C. V., Linc C. I.: "Development of a patent document classification and search platform using a back-propagation network", Expert Systems with Applications, Vol. 31 (4), November 2006, pp. 755-765.
12. Goldfire Innovator, www.invention-machine.com.
13. Salton, G., McGill, M. J.: "Introduction to modern information retrieval", McGraw-Hill, ISBN 0070544840, 1983.
14. Krippendorf K: "Content Analysis: An Introduction to Its Methodology". Thousand Oaks, CA, Sage 2004.
15. Cascini G., Fantechi A., Spinicci E.: "Patent Clustering through selected functions", submitted for publication to the Journal of Information Processing & Management, 2007.
16. Hui B., Yub E.: "Extracting conceptual relationships from specialized documents", Data & Knowledge Engineering, Vol. 54 (1), July 2005, Pp. 29-55.

# Automatic shape variations for optimization and innovation

*Shape Optimization of Cylinderhead Gasket using CFD*

Noel Leon, Jose Cueva, Cesar Villarreal, Sergio Hutron, German Campero
Tecnologico de Monterrey, CIDYT-CIII
Ave. Eugenio Garza Sada # 2501
Col. Tecnológico, Monterrey, NL, CP. 64839, Mexico
http://cidyt.mty.itesm.mx

**Abstract**. This paper presents an innovative tool and method that allow efficient innovation of shape and topology of virtual parts at both mesh and CAD levels using optimization methods. The method consists of automatic variations of shapes in CAD/CAE environments that allow effective search for new shapes that are not considered initially by designers.

## 1 Introduction

Contemporary designers face the dilemma of doing design tasks in a context where the available tools and methods are not always adapted to satisfy design requirements that are increasingly more focused on potential creativity enhancement [1].

This paper addresses the main motivations of the industrial sector, regarding the engineering innovation activity with computer tools and methods. It is focused on the innovation of shape and topology using a proven and mature method that has been used successfully for parametric optimization. In addition the method extends its application to design automation and CAI for integrating these methods and tools into engineering processes.

It is known that, commonly, performance improvement is first obtained through quantitative changes in parametric design in search for optimization, but once the resulting improvement reaches its limit and further performance improvements are not possible, new searches must be carried on through qualitative changes or paradigm shifts that lead to innovation [2]. Optimization, then, is a search process that looks for the most advantageous state of equilibrium before a contradictory situation where the improvement of one or several performance characteristics,

*Please use the following format when citing this chapter:*

Leon, N., Cueva, J., Villarreal, C., Hutron, S., Campero, G., 2007, in IFIP International Federation for Information Processing, Volume 250, Trends in Computer Aided Innovation, ed. León-Rovira, N., (Boston: Springer), pp. 179-188.

deteriorates others. In order to get out of this deceptive goal, innovative concepts are needed by changing not only parametric values, but also shapes, topologies and physical principles.

Currently available CAD and CAE systems were originally conceived for facilitating only parametric variations on modeled parts. Parametric modeling has simplified the design process because it allows easy modification of parts. In recent times, topology optimization methods have been also introduced in meshing environments to improve product performance [1, 2]. These topological optimization functions are currently used to find optimum topologies and shapes for given parts under prearranged conditions. This is achieved by describing a defined space for the part through a Finite Element (FE) mesh model while an optimization algorithm finds an optimal material distribution within a series of established restrictions. Properties of the FE model such as density or Young modulus are modified during the optimization process until an optimum shape is obtained [3, 4]. This type of mesh-level variations is practical for finding suggestions regarding part shapes and topologies, however shapes and topologies obtained this way are not models with a CAD structure, and they require manual post processing or even a complete redesign if it is desired to convert them into a full structured CAD model [5].

In this paper innovative tools and methods are presented that allow efficient optimization and innovation of shape and topology of virtual parts at both mesh and CAD levels.

In this work new advances are shown for implementing methods that may be integrated into conventional CAD/CAE systems for executing shape and topology changes that transcend parametric values while searching for performance enhancements with the aid of genetic algorithms and shape and topology variations at both CAD and mesh models that may be converted in both directions.

## 2 Splinization and Mesh Morphing

For the development of this method and its adaptation to the current CFD analysis methodologies the splinization approach presented in the past CAI Conference [6] needed to be translated into the CAE domain. This paper describes how the splinization concept was translated into the meshing software for integrating the shape variation method with the CAE analysis method.

Even though the main focus on the splinization technique is about shape variations that allow the persistence of CAD models, a new interesting way of obtaining these same advantages is being developed to allow special shape variations at mesh level and then be translated into direct modification for a previously "splinized" CAD feature. For this purpose mesh morphing in existing commercial finite element meshing software was used, which allow shape variations to be made to a mesh model without remeshing it. One further reason mesh morphing becomes advantageous in this context is because the interpolation of previous results can save a lot of computational time in CFD simulations. Since the changes in geometry are small compared to the size of the complete model, previously converged simulation result (or interpolated new result) can be used as an initial guess and speed up the simulation process. To take an advantage of converged solution, it is good to have a mesh model that remains consistent in its number of elements and elements' IDs.

Since any modification that requires remeshing would cause inconsistency in these matters, mesh morphing is a good option. In mesh morphing, a domain is an area where shape modification propagates to define their movements. The way the mesh surrounding the shape is modified depends on a bias value (so called a "handle") that is equivalent to the tension of a Bezier curve. Within each domain handles are defined.

Under these circumstances, a direct relation can be established between these morphing handles, and the position of the control points that define a spline. With this, a connection between the splinization approach and mesh morphing has been established. A morphing handle approach reduces the number of design variables while splinization increases degree of freedom in shape. The final goal is to integrate the CAD/mesh shape optimization process through splinization methodologies with finite element meshing software for CFD and structural simulations and genetic algorithms in an automatic manner.

## 2.1. Mesh morphing

As known from finite element method, a mesh is constituted of elements, which are constituted of nodes and edges connecting two nodes. Nodes are defined by coordinates in space which, by being strategically modified, can alter the entire mesh model. By displacing every set of node coordinates, the whole mesh model can be moved. In the same manner, it can be rotated, scaled and stretched in every way.

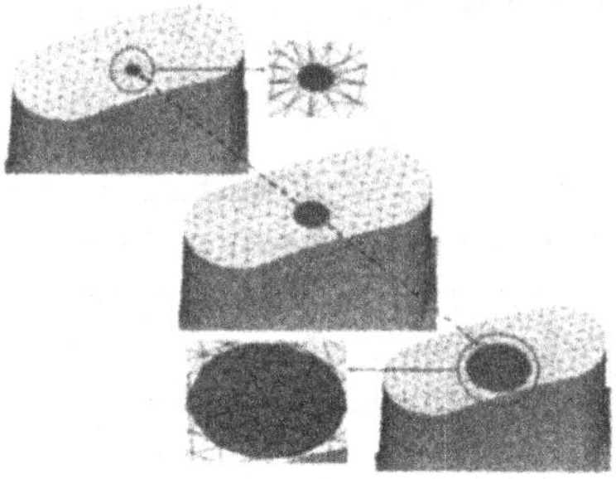

**Fig. 1.** Extensive morphing diminishes element quality.

Because it is sometimes desirable to modify only a certain feature within a mesh model, manipulating node coordinates must be done in a very controlled manner to maintain model consistency and quality of the mesh elements. In fact, the best way to modify a certain shape would be to alter the nodes defining it, and then smoothing the node displacements so that the perturbation of the elements is minimized and mesh topology is preserved. This effect is known as mesh morphing, which is a tool included in some meshing software.

For each modifiable shape, edge domains are formed in the shape that defines the mesh boundaries where morphing may be performed without significant deterioration of the mesh quality of the region. Several stepped regions allow defining a bigger region than would be allowed by mesh quality criteria when performing the variation in a big region. These stepped regions are placed along the curve of the edge domain that will be modified individually.

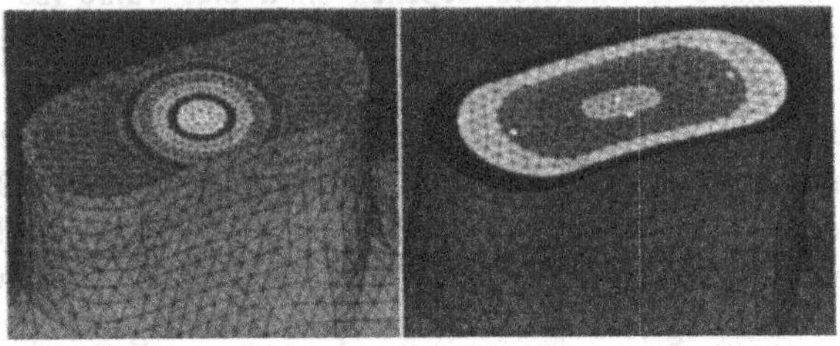

Fig. 2. Mesh constructed with proper pre defined morphing levels.

With this tool, a domain can be created to define a group of elements to be taken into account as a morphing unit. Within a domain, handles allow the actual node manipulation. Handles can define curvatures or approximate diameters in the case of circular domains; they can be also freely moved, anchored at a given point, assigned movement restrictions to, etc. Even though this kind of variations have major limitations because of mesh quality constraints, they are very practical when small changes in shapes need to be made.

## 3. Case Study: Shape Optimization of CylinderHead Gasket using CFD

The traditional process involves geometry, meshing tools and Computational Fluid Dynamics (CFD) simulations for the performance enhancement of the cooling water jacket of cylinder engines. This process has been usually performed manually modifying the gasket holes dimensions based on the results of CFD simulations subject to manufacturing, packaging and material constraints. However, this process is highly time consuming.

Water jacket is designed for peak power condition that is representative of the worst case scenario with respect to heat rejection to coolant with the flow rate specified at peak pump output.

For designing the water jacket, predominant design variables are number, dimension and location of each gasket hole. Gasket holes should reside inside of intersection of block and head water jackets.

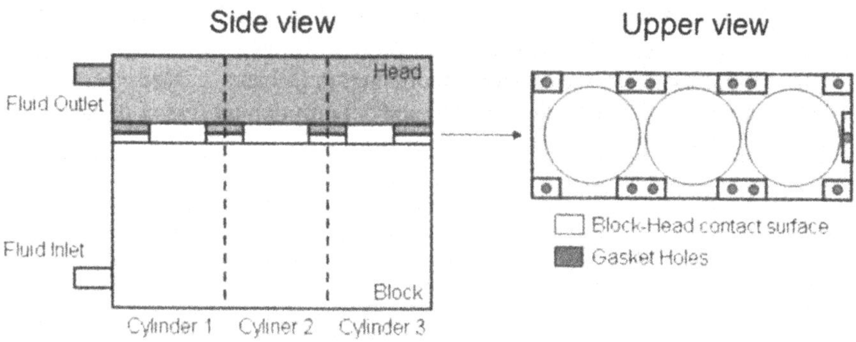


**Fig. 3.** Block and head water jackets intersection and gasket positioning.

Currently, design iterations take about 1/2 week for each change to do the mesh generation and run the analysis. The simulation time for each simulation using a 4CPU machine is about 16-20 hours. Under these conditions the possibilities of variations of shapes and topologies are very limited and therefore only changes of the holes diameters are commonly introduced.

The goal of applying shape optimization concepts and methods is to significantly reduce the time for finishing a design, by providing a functional modeling methodology that integrates the engineering concept and the geometry restricted to manufacturing, packaging and material constraints. Additionally performance increments beyond those achieved by traditional manual methods may be expected as not only changes in diameter values but also shape and topology variations

The design has to be performed under the restrictions coolant flow distribution around each cylinder, flow-distribution near critical areas such as exhaust valve bridge, spark plug and peak velocity zones that are prone to metal erosion. Thermal analyses are performed to predict heat transfer coefficients and for transferring the information to structural analysis. Thermal analyses provide the coolant flow rate from the water pump performance, heat fluxes corresponding to the peak power condition in different regions of the water jacket.

### 3.1 Optimization of the engine cooling process

Holes in a cylinder head gasket are the only available passage for cooling fluid between the cylinder head and cylinder block. The pattern of gasket holes defines the flow conditions within the water jacket that allows heat exchange from cylinder head and block to coolant. As shown in [7], changes in gasket holes pattern can positively impact heat transfer coefficients for the cylinders, thus improving engine cooling. In this case the objective of gasket holes optimization is to maximize the minimum area averaged heat transfer coefficient as well as to minimize the maximum difference among heat transfer coefficients among cylinders.

The overall process consists of software applications to run DOE, mesh update, CFD analysis, and Process integration for performance optimization.

With the help of the mesh morphing process explained earlier, in the DOE step re-meshing is avoided while boundary conditions are changed.

The optimization step uses mesh morphing to achieve a new mesh in a reasonable time. Overall procedure of DOE and optimization process of gasket holes is presented in Figure 4.

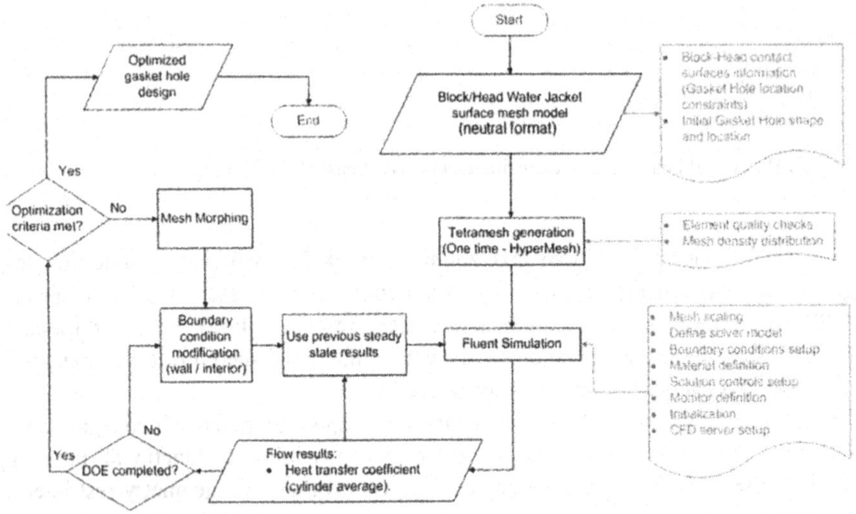


**Fig. 4.** Overall procedure of DOE and optimization process of gasket holes.

The main benefit of this process template is reduction of man-hours required for set-up and post processing by eliminating routine tasks that are now implemented automatically.

After this mesh-based shape modification technique reaches a final design; work would be still required in order to feed this final design back to the designers. The implemented procedure allows closing this shape optimization process by sharing of equally flexible parameterization techniques both in the CAD and CFD environment by feeding the shape variations of the mesh morphing controlled by the handles as shown in Figure. 5 to the CAD environment as control points of a spline where the final values of the morphing handles become the control points of the spline. Through this approach, the mesh-based shape optimization can be performed and then, when the optimal shape design has been defined, the final shapes can be automatically translated to the CAD design.

### 3.2 Preliminary results

At this point the investigation has been conducted in a simplified model shown in the Figure 6, in which six holes (shown in Figure 7) perform a scale factor, without shape and/or topology variations yet

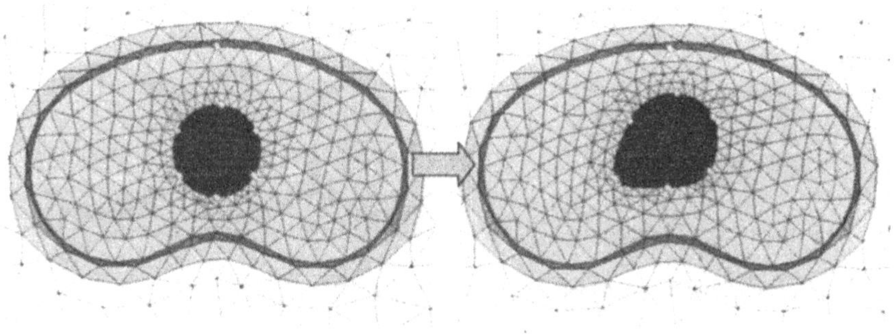

**Fig. 5. Gasket hole modification with splines.**

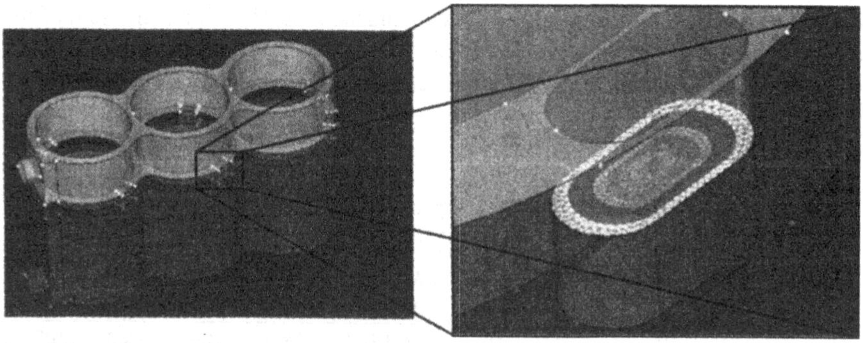

**Fig. 6.** Simplified model gasket holes.

The mesh dfined contains small elements (0.25 mm) where is necessary to be more precise for example holes and thin walls. Larger elements (up to 3 mm) are used in other areas. As boundary conditions an initial coolant temperature of 366°K was defined. The coolant properties were defined based on a common mixture of water and ethylenglicol (r=1028.5 and m=0.00072). The neighborhood cultivation algorithm was used with the criteria of maximum 400 iterations.

The process integration for performance optimization in the software applications has been already achieved. As mentioned in the previous chapter, process integration helps automate the mesh modification and CFD analysis. Therefore it is possible to perform a DOE and/or to conduct an optimization using a search algorithm such as a genetic algorithm.

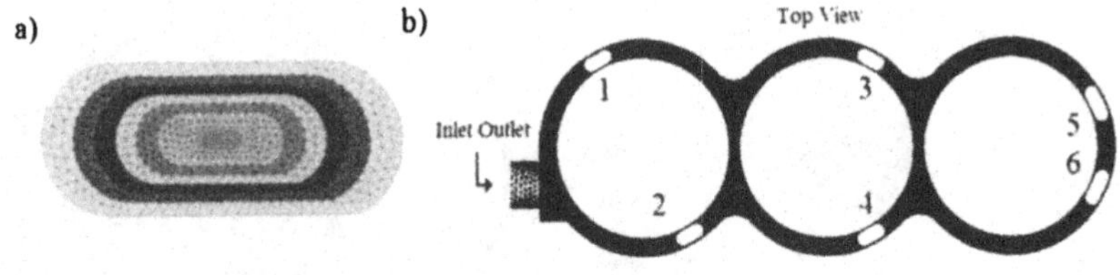


**Fig. 7. a) Gasket hole mesh, b) Gasket holes positioning.**

With the process integration a fractional factorial DOE analysis with four levels for each hole was performed. These hole levels are classified into a) fully opened; b) one third of the total area; c) two thirds of the total area; and d) completely closed.

As explained earlier, the objective is to maximize the minimum heat transfer coefficient (HTC) as well as to minimize the maximum difference of HTC among cylinders. To perform the DOE analysis the objective function to maximize is defined as follows: HTC sum of the three cylinders minus four times the sum of the differences among the cylinders. The results of the DOE analysis are shown in Table 1 and the Pareto frontier for visual approach of the optimization can be seen on the Figure 8.

**Table 1.** Fractional Factorial DOE analysis top ten HTC.

| Rank | Gasket Hole Area (Percentage) | | | | | | HTC Cylinders ($W/m^2K$) | Differences between Cylinders ($W/m^2K$) | | | Objective |
|---|---|---|---|---|---|---|---|---|---|---|---|
| | 1 | 2 | 3 | 4 | 5 | 6 | 1+2+3 | \|1-2\| | \|2-3\| | \|3-1\| | |
| 1 | 1/3 | 0 | 0 | 1/3 | 1/3 | 1/3 | 271.9 | 28.2 | 8.9 | 19.2 | 103.0 |
| 2 | 0 | 1/3 | 1/3 | 0 | 1/3 | 1/3 | 269.4 | 28.0 | 9.2 | 18.8 | 101.4 |
| 3 | 0 | 0 | 1/3 | 1/3 | 1 | 1 | 233.9 | 24.3 | 0.0 | 24.3 | 88.1 |
| 4 | 1/3 | 1/3 | 0 | 2/3 | 2/3 | 2/3 | 246.3 | 26.4 | 1.5 | 24.9 | 88.0 |
| 5 | 1/3 | 1/3 | 2/3 | 0 | 2/3 | 2/3 | 246.5 | 27.9 | 2.1 | 25.7 | 79.2 |
| 6 | 0 | 0 | 1 | 1/3 | 2/3 | 2/3 | 238.2 | 20.2 | 6.3 | 26.5 | 79.0 |
| 7 | 0 | 0 | 1/3 | 1 | 2/3 | 2/3 | 222.6 | 17.9 | 9.9 | 27.7 | 56.2 |
| 8 | 2/3 | 0 | 1/3 | 2/3 | 2/3 | 1/3 | 244.1 | 27.0 | 5.1 | 32.1 | 51.6 |
| 9 | 1/3 | 1/3 | 2/3 | 2/3 | 1 | 1 | 233.1 | 25.5 | 5.0 | 30.5 | 50.1 |
| 10 | 0 | 2/3 | 2/3 | 1/3 | 1/3 | 1 | 234.2 | 28.3 | 2.9 | 31.2 | 47.1 |

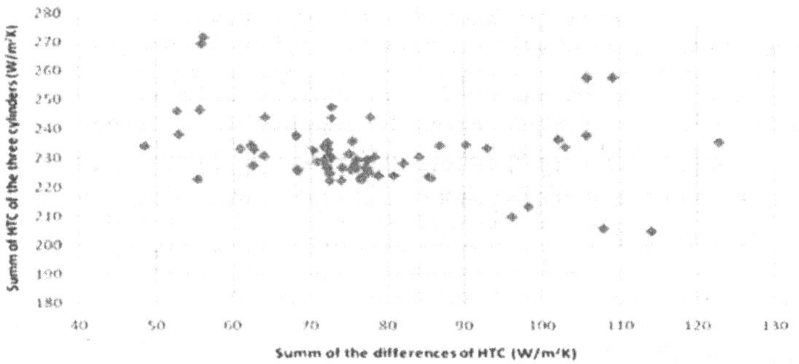


**Fig. 8.** Pareto frontier between the two objectives.

In these preliminary results, the tendency of having bigger holes while they are farther from the inlet is higher. Based on this fact, levels should be chosen for the gasket holes as a starting point so that optimum can be reached quickly.

## 4. Future work

Taking advantage of the automatic shape variation presented in this paper, it is proposed to apply this technique looking for innovative shapes of Savonius tye rotors for wind generators. The Savonius wind turbine was developed in the early 20's of last century and has been undermined because of its low efficiency in comparison with traditional wind turbines of horizontal axis. Later groups of investigators have worked finding new shapes of the Savonius rotor making subtle changes to its profile shape, but the traditional geometrical features such as arcs and lines have remain unchanged. By using splines in the rotors design it is possible to create forms never explored before, which may be analyzed trough the techniques described in this paper for the Shape Optimization of CylinderHead Gasket using CFD.

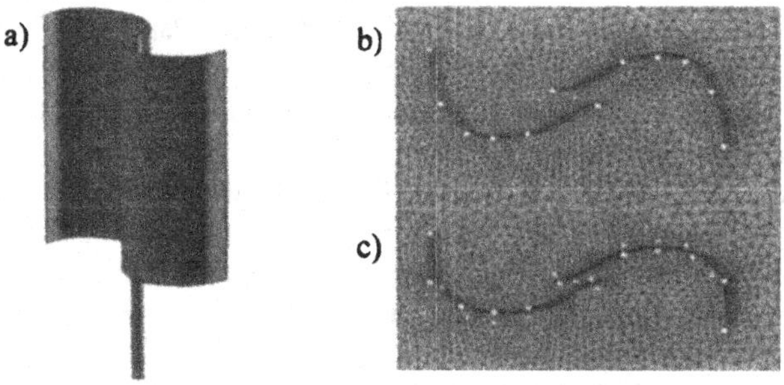


**Fig. 9.** a) Traditional Savonius rotor, b) original profile shape, b) modified profile shape using splines.

## 5. Conclusions

Computer Aided Innovation may benefits from mature optimization methods for performing shape and topologic variations in CAD/CAE environments leading to innovative shapes and topologies that allow achieving higher performance when parametric optimization faces technical contradictions. Using "splinization" techniques together with mesh morphing and genetic algorithms facilitates the search for higher performance exploring innovative shapes that avoid expensive manual trial and search methods.

## 6. Acknowledgments

The authors acknowledge the support received from Tecnológico de Monterrey through Grant number CAT043 to carry out the research reported in this paper. Authors acknowledge collaboration effort with Vamshi Korivi, Surendra Gaikwad and Su Cho working from Aerothermal Center of Competence and Product Development IT, Chrysler LLC.

## References

1. Cavallucci D, Leon N, Towards "Inventiveness-Oriented" CAI Tools, Proceeding of IFIP 18th World Computer Congress, Topical Sessions, Building the Information Society; pp. 441-452, Toulouse, France , 22-27 August 2004.
2. Leon N, Gutierrez J, Martinez O, Castillo C, Optimization vs. Innovation in a CAE Environment, Towards a "Computer Aided Inventing" Environment. Proceeding of IFIP 18th World Computer Congress, Topical Sessions, Building the Information Society; pp. 487-495, Toulouse, France , 22-27 August 2004.
3. Pandley MD, and Sherbourne AN, Mechanic of shape Optimization in Plate Buckling, Journal of Engineering Mechanics, Vol. 118, No. 6, pp.1249-1266, June 1992.
4. Vicini A and Quaglierella D, Airfoil and wing design through hybrid optimization strategies, 16th Applied Aerodynamics Conference, American Institute of Aeronautics and Astronautics, 1998, AIA Paper 98-2729.
5. Binder T, Hougardy P, Haffner P, Optimierung von Guss- und Schmiedeteilen bei Audi, Audi AG, D-85045 Ingolstadt, Audi AG, D-74172 Neckarsulm, FEM, CFD, und MKS Simulation, 2/2003.
6. León Rovira, Noel; Silva Sierra, David Alejandro; Cuevas, José María; Gutierrez Hernández, Jorge César, Automatic Changes in Topology of Parts and Assemblies in 3D CAD systems. Proceeding of the 1st IFIP Working Conference on Computer Aided Innovation, Ulm, Germany pp: 96-107, November 2005.
7. Shih S, Itano E, Xin J, Kawamoto M and Maeda Y, Engine Knock Toughness Improvement Through Water Jacket Optimization, SAE paper 01-3259, SAE International, 2003.

# Enhancing interoperability in the design process, the PROSIT approach

Umberto Cugini[1], Gaetano Cascini[2], Marco Ugolotti[1]
[1] Politecnico di Milano, Dipartimento di Meccanica, Milano, Italy
{umberto.cugini, marco.ugolotti}@polimi.it
[2] Università di Firenze, Dip. di Meccanica e Tecnlogie Ind.li, Florence, Italy, gaetano.cascini@unifi.it

**Abstract.** The paper presents a methodology developed within the PROSIT project, aimed at the improvement of the product development cycle through the integration of Computer-Aided Innovation systems with Optimization and PLM/EKM systems. The interoperability of these tools is obtained through the adoption of Optimization systems as design analysis means and the definition of formalized and validated procedures and guidelines. The logic of the proposed methodology is explained through a detailed study case related to the design of a plastic wheel for light moto-scooters.

## 1 Introduction

The necessity to succeed in the global market or, at least, to survive in such a competitive environment, forces industries to systematically innovate products, to reduce their costs and to introduce them faster.

The achievement of this goal has led to the introduction of new technologies and new working approaches in the product development domain in order to improve (often in a very radical way) design activity performances. In the conceptual design phase, for instance, several methods and tools have been developed over the years to support the systematic transfer of innovative solutions among different technical fields. It is the case of Computer Aided Innovation (CAI) systems, which help engineers and technicians addressing design problems and guiding them to new possible solving approaches [1]. Moreover, methods for structural and topological optimization (TO), based on the use of a generative algorithm, are actually used by a lot of practitioners to obtain optimal geometrical solutions [2]. Knowledge-Based Engineering systems (KBE) finally, support designers' activity through rules and knowledge re-use, thus reducing the product development time without effecting its functionality, quality and security [3]. These tools have demonstrated their relevant potentialities to increase the effectiveness of specific design activities where they are

*Please use the following format when citing this chapter:*

Cugini, U., Cascini, G., Ugolotti, M., 2007, in IFIP International Federation for Information Processing, Volume 250, Trends in Computer Aided Innovation, ed. León-Rovira, N., (Boston: Springer), pp. 189-199.

used. Technological potentials of these systems, however, especially the potential od a smooth and effective integration, have not been fully exploited and it is mainly due, on one side, to a poor integration between different applications, and on the other side by the difficulty to fully understand how to take advantages of these tools and how to effectively use them in the context of the product development process. Indeed, it has been estimated that the United States industry spends billions of dollars as a result of poor interoperability between computer-aided engineering software tools [4].

A full integration of these technologies is still far to be reached [5], and a big effort is required to set up successful collaborations and to push companies to focus their attention on the adoption of new organizational paradigms to better coordinate the design activity in such a context. The capability to support in a more integrated way all the stages of the product development process will be one of the most important competitive factors for these systems in the next years. In a concurrent engineering view, in fact, it is required that the phases of conceptual design, optimization and detailed design would be integrated as far as possible (Fig. 1). In order to reduce development time and increase activities' effectiveness, design and product development process have to be considered as a continuous iteration among these phases.

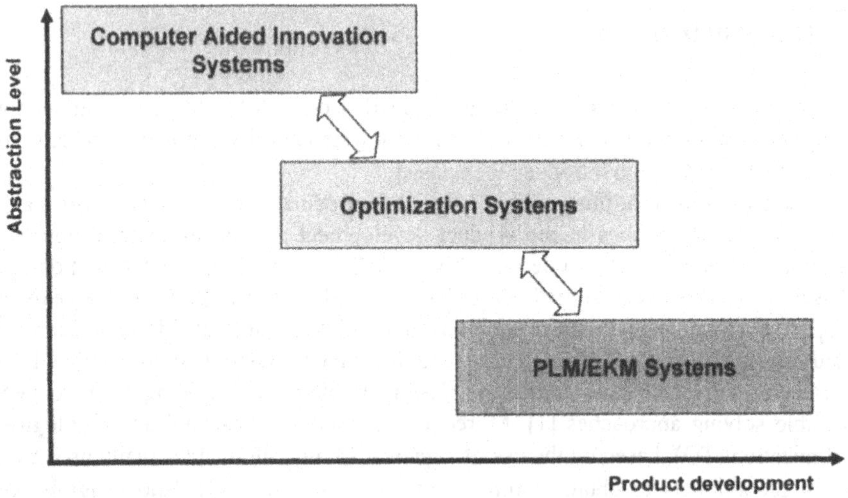


**Fig. 1.** Methods and systems to support product development process

Real potentialities of CAI, TO and KBE tools are still far to be reached. Lack of integration between different systems has reduced the impact of these tools in the design process. One of the main reasons is that these applications have been developed in the past mainly as standalone solutions, as islands of automation, unable in most cases to communicate and interoperate each other. The development of a more interoperable design process, in fact, is not only a matter of new advanced and independent IT solutions, but it has to be deeply focalized in the development of methodologies and guidelines aiming at supporting design teams in a more effective and efficient way, improving the advantages provided by the adoption of available product development systems, data management activities and knowledge sharing among design teams.

In the last years several authors have tried to integrate optimization techniques in product design processes with classical approaches such as feature-recognition [6] and design-by-feature [7]. Innovative frameworks have been also proposed to introduce optimization processes from the conceptual design stage, for instance by using hybrid CAD/FEM models built up by predefined components with predefined structural characteristics [8]. Other attempts have been tested, aimed at integrating optimization and other design activities through iterative approaches where the initial geometry is submitted to a Geometry (CAD) -> Simulation (FEM) -> Optimization -> Geometry (CAD) cycle [9]. However the proposed solutions implemented so far did not result in an integrated approach, neither have been expressed into formalized procedures for the systematic introduction of structural optimization systems in the design process.

Finally, the introduction of innovations into a process requires a preliminary study to estimate the impact of the proposed changes, before making the required investment to introduce them. It is thus necessary to adopt a systematic approach to the definition of a process modification through an effective and efficient introduction of technologies, and to support new technologies evaluation and impact introduction analysis, considering aspects as costs, times, design errors and iteration reduction, etc. The authors have already developed in the past some evaluation metrics for the assessment of the benefits provided by the adoption of new technologies in a product development process as a comparison between its As-Is and To-Be models [10].

This paper tries to put in evidence the meaning of interoperability in the design process and to present a methodology developed by the authors for the design and implementation of more interoperable design environments. The methodology here presented has been developed in the frame of an Italian ministry co-financed project PRIN 2005, named PROSIT (From Systematic Innovation to Integrated Product Development).

The main objective of this project is to study and test possible solutions for integrating innovative tools such as PLM (Product Lifecycle Management) and EKM (Engineering Knowledge Management), with CAI systems and TO tools within the product development cycle. The rationale of the proposed research is the lack of formalized and validated procedures allowing the systematic introduction and integration of these tools in the design process.

## 2 Methodology

According to the diagram of Fig. 1, the PROSIT project aims at bridging three different classes of product development methods and systems, CAI and Optimization systems from one side, Optimization systems and PLM/EKM tools to the other.

The main idea of the methodology developed in the frame of the project to link CAI and Optimization systems is the adoption of the latter tools not just to generate optimized solutions, but also as a design analysis tool, capable to outline critical aspects of a mechanical component in terms of conflicting design requirements or parameters.

The logic behind CAI systems is mostly related to the TRIZ theory, i.e. to the refusal of trade-offs; thus, they are apparently in conflict with the logic of optimization, seen as minimization of negative issues within a given set of constraints. Nevertheless, as explained below, optimization systems can be used in a novel mode, such that they can play a relevant role in the identification of contradictions.

More specifically, the traditional approach to optimization involves the application of a complete system of constraints and loads to the geometry for describing all the design requirements.

It is worth to notice that this "optimal" i.e. "best compromise" solution is unnecessarily satisfying. It's often useful, before moving towards the detailed definition of the product architecture, to re-discuss already made assumptions, in order to obtain a solution which better satisfies general system objectives. On the basis of these considerations, the authors propose to perform a set of mono-objective optimization tasks in order to put in evidence conflicts among geometrical elements of the system under analysis.

Therefore, the work presented in this paper has been focalized on the definition of guidelines to make systematic the translation of a system functional model and its design requirements into a set of mono-objective optimization problems, aiming at satisfying a single requirement at one time. These guidelines have to be intended as a help for designer when they are asked to translate the functional architecture obtained as output of CAI system into a set of design variables and a set of constraints and load conditions to be optimized separately.

### 2.1 From optimization analysis to geometrical contradictions identification

The rationale behind the adoption of Optimization systems as a means for design analysis is the following:

- defining a single multi-goal optimization problem leads to a compromise solution;
- besides, defining N complementary mono-goal optimization problems, each with specific boundary conditions, leads to N different solutions;
- these solutions can be conflicting and this is the key to find contradictions.

According to this statement, the PROSIT design flow is structured as depicted in Fig. 2. The process starts with the definition of a set of single-goal optimization tasks, each representing a specific operating condition and/or a given design requirement. If each output solution satisfies the design objectives and they mutually fit each other, the process doesn't require any iteration and a detailed CAD model can be produced: the definition of a bridge between Optimization and PLM systems is a further goal of the PROSIT project, but it will be briefly presented below.

Besides, if the solution of at least one of the optimization tasks doesn't fit the design requirements and/or the optimization tasks lead to conflicting geometries, the system must be further investigated in order to extract the geometrical contradictions. Geometrical contradiction concept is based onto studies published by Vikentiev about Geometrical Effects (GE). Somehow it can be stated that "GE start where physical and chemical effects end", or more precisely, unlike chemical effects, which enable to obtain some substances from others by the absorption or isolation of energy, and physical effects that enable to transform one form of energy into another, GE usually organize and redistribute flows of energy and substances that are already available in the system. In considered context, talking about design embodiment, functional architecture of system is already defined and then the work is essentially focused on geometrical field.

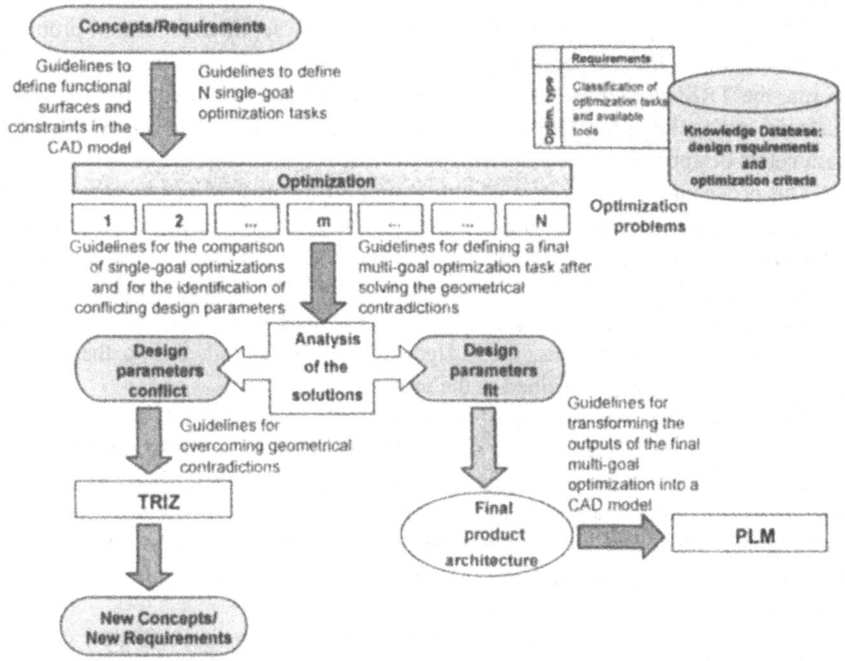


**Fig. 2.** Design flow according to the PROSIT approach

More specifically, according to the differences between the results of the single-goal optimization tasks and to the nature of the conflicting design parameters, the geometrical contradictions can be classified into:

- 1D/2D/3D size contradictions: a linear dimension (e.g. length, width, height, thickness etc) should be high and should be small; a surface/volume should be big and should be small etc.
- shape contradictions: an element or a detail should assume different forms (e.g. sharp and rounded, circular and polygonal etc);

- topological contradictions: an element or a detail should assume different topologies and/or orientations (e.g. monolithic and segmented, horizontal and vertical etc.).

A detailed description of the nature of these contradictions and the strategies to overcome them is out of the scopes of the present paper. Interested readers can find further details in [11].

Basically the methodology guides the designer according to the classical TRIZ approach. The application of TRIZ instruments allows, in synthesis to: identify the operational zone within the design space where the geometrical contradiction resides, check if contradictory requirements for the design parameters co-exist in the operational zone and time, check if those contradictory requirements co-exist under any condition and evaluate the opportunity to overcome them by means of a separation strategy or through a transition to subsystem elements or to its supersystem.

Closing the loop, as a result of this activity, a new set of optimization problems can be identified and can be solved making use of the optimization tools. In other words, the TRIZ principles are used to redefine the design volume, the functional surfaces and/or the optimization constraints so that the conflict between the design parameters disappears.

This procedure has to be iterated until optimization process' results converge, i.e. the geometries generated by the different single-goal optimization tasks fit each other. The final output of the Optimization system must be translated into a complete feature-based CAD model. It is worth to note that despite several features recognition systems are actually available on the market as additional modules of more advanced PLM systems, those algorithms typically fail, due to the intrinsic complexity of the shapes obtained in the previous step.

Thus, the proposed approach consists in the definition of a best practice implemented in a Knowledge Base user interface, which supports the designer in the translation of the output of a design optimization into a feature-based model, with simplified and possibly automated steps. Therefore, the project is oriented to capture the designer intent and knowledge during the optimization process through a KB interface capable to automate, support, and simplify geometric model rebuilding operations.

The development of the Knowledge Base relies on two main postulates:

- whatever is the mechanical part to be modeled, it is possible to segment the model into a set of invariants "typical" volumes, on the base of the product and design requirements;
- these "typical" volumes, characterized by a specific topology and some relations among geometric entities, can be adopted as seeds for the generation of the feature-based model after the optimization phase.

Further details about this work-package of the project will be published in the next future.

# 3 Study Cases

The methodology here presented is currently under validation and it has been applied in several study cases in order to verify its feasibility, its benefits and drawbacks, in terms of time, quality and costs improvements, and in terms of capability to establish more collaborative working approaches. Five main fields of interest have been identified and they concern the re-design of the following mechanical systems:

- a scooter wheel using polyamide;
- a connecting rod for racing internal combustion engines;
- a hinge for glass doors;
- a hinge for chest freezers;
- a climbing hook.

Due to space limitations just the first design task will be detailed in order to clarify the PROSIT methodology with an exhaustive example.

## 3.1 Re-design of a scooter wheel

This test case has been inspired by a real case study developed during a collaboration of the authors with the Italian motorbike producer Piaggio [12]. The goal of the project was the design of a plastic wheel for light moto-scooters mainly aimed at costs reduction, of course without compromising safety and mechanical performances.

The traditional approach in Piaggio to assess the conformity of a wheel consists in three different experimental tests:

- deformation energy under high radial loads/displacements (simulating an impact against an obstacle);
- fatigue strength under rotary bending loads (simulating the operating conditions such as curves);
- fatigue strength under alternate torsional loads (simulating the accelerations and decelerations).

These tests have been adopted as reference criteria for design optimization, under the constraint of manufacturability through die-molding and the goal of minimizing mass since this parameter is directly related to costs.

Two functional surfaces were identified: the hub and the rim. The design domain for the topological optimization task coincides with the envelope of the volume of a classical wheel, while the internal side of the rim and the hub are assumed as invariant.

According to the proposed methodology, the first step consisted in the definition of three different mono-requirement topological optimizations related to the above mentioned complementary tests. As a result, two different mass distributions were generated, the first related to the radial and torsional tests, the second to the rotary bending test (Fig. 3): more specifically, the first geometry suggests the creation of a flat web between the hub and the rim; while the second leads to a number of radial spokes with transversal ribs. Before than facing such a topological contradiction, a further investigation must be done because the analysis revealed that the radial test by itself didn't meet the expected requirements.

When a mono-goal optimization doesn't converge to any acceptable solutions, the PROSIT guidelines suggest to split the problem further and to define a subset of optimization tasks with the same objective function, by removing the optimization constraints one-by-one.

As a result the user is driven to the identification of design constraints in the following form: the constraint X should not be respected in order to fulfill the goal G, but it should be respected in order to satisfy the requirement R.

In this case, the constraint to be removed in order to meet the design goal is the draw direction for manufacturability issues. In facts, the elimination of such a constraint leads to the geometry shown in Fig. 4: a hollow wheel with a double web supporting the side of the rim. Hence the contradiction can be expressed in the form: "the wheel should present an axial draw direction in order to preserve manufacturability and should not present an axial draw direction in order to provide a sufficient radial stiffness".

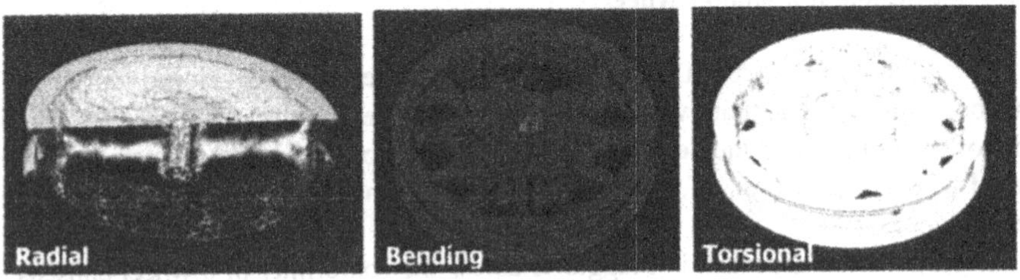


**Fig. 3.** Re-design of a scooter wheel: first step, comparison of the geometries arising from three complementary single-goal optimization tasks.

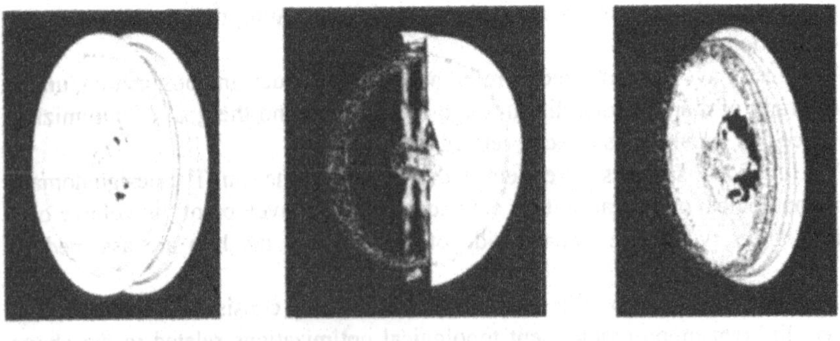

**Fig. 4.** Re-design of a scooter wheel: second step, in order to meet the design objectives, the manufacturability constraint should be removed.

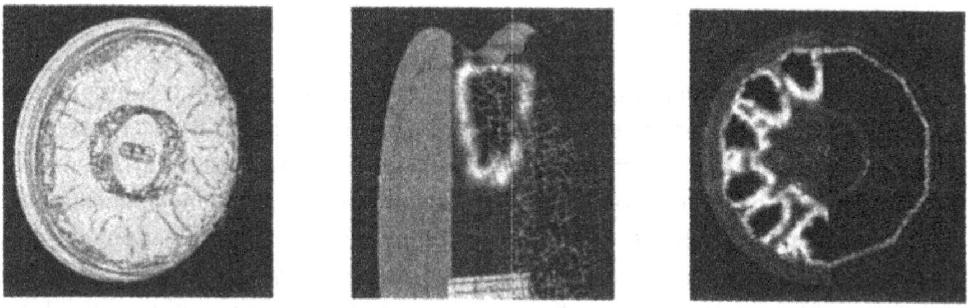

**Fig. 5.** Re-design of a scooter wheel: the geometrical contradiction can be overcome by applying the segmentation principle.

The guidelines extracted from the TRIZ instruments for overcoming the geometrical contradictions suggest, among the others, the application of the segmentation principle. In other words, the design space can be divided in two parts, so that the manufacturability is preserved when the two parts are separated, while the stiffness satisfies the requirements when assembled. Fig. 5 shows the results of the topological optimization obtained by dividing the wheel into two halves.

It is worth to note that the web is characterized by a number of transversal ribs that remind the spokes suggested by the optimization under rotary bending loads (Fig. 3). Therefore, in this case the solution of the contradiction within the radial load test brings to a geometry that fits with the results of the other mono-goal optimization tasks above defined and the overall process is converged.

This conceptual result can be translated into a geometrical model to be defined through technological features according to the second main goal of the PROSIT project. A standard FEM analysis of the final geometry revealed that a segmented plastic wheel made with a resin PA6 with a 30% volume content of glass fibers, (modulus of elasticity about 16 GPa and stress to failure about 230 MPa) is capable to pass all the virtual tests required by Piaggio proprietary standards.

## 4 Conclusions and Future Works

The present paper has addressed the integration of Computer-Aided Innovation systems, Optimization systems and PLM/EKM tools as a means to improve the innovation resources and the efficiency of a product development cycle. The rationale of this research is the lack of formalized and validated procedures allowing the systematic introduction and integration of these tools in the design process. The development of these integration guidelines, that is the most innovative aspect of the research, should provide advantages in terms of design costs reduction, errors reduction, product quality improvement, process execution time and more effective internal and external knowledge use and share.

A relevant aspect of the PROSIT project is the integration of apparently incompatible tools, thanks to the new role and way of usage of the Optimization Systems.

Moreover, in the mind of the authors, the methodology, through the definition of guidelines and best practices, can help overcoming classical interoperability problems affecting the design process, providing a common base for knowledge sharing and for a better interconnection between different systems and applications.

It is worth to notice that the PROSIT project doesn't aim at the creation of a fully automatic system for design embodiment, because both the comprehension of the root-cause of a geometrical contradiction and, most of all, the translation of the TRIZ principles into a new set of optimization tasks, requires a creative even if systematic step, demanded to the designer. Besides, the results obtained so far suggest the investigation of semi-automatic procedures to speed-up some routinary tasks like the comparison of the outputs of the single-goal optimization tasks as well as the removal of the optimization constraints one-by-one, when a single optimization doesn't converge to a solution.

## 5 Acknowledgements

The PROSIT project (Product development and systematic innovation www.kaemart.it/prosit) is co-funded by the Italian Ministry of University and Research.

The authors would like to thank M. Bordegoni e M. Bertoni from Politecnico di Milano and F. Rotini and F. Frillici from Università di Firenze, M. Muzzupappa from Università della Calabria and F. Cappello from Università di Palermo for their contribution to the development of the PROSIT project.

## 6 References

1. G. Cascini: "State-of-the-Art and trends of Computer-Aided Innovation tools - Towards the integration within the Product Development Cycle", Building the Information Society, Kluwer Academic Publishers (ISBN 1-4020-8156-1), 2004, pp. 461-470.
2. M.P. Bendsoe, O. Sigmund: "Topology Optimization - Theory, Methods and Applications", Springer, 2003.
3. M. Bordegoni, U. Cugini: "Knowledge-driven method for integrated design", Proceedings of the International CIRP Design Seminar, Grenoble, France, May 12-14, 2003.
4. Simon Szykman, Steven J. Fenvesa, Walid Keirouzb, Steven B. Shooter: "A foundation for interoperability in next-generation product development systems", Computer-Aided Design, Volume 33, Issue 7, June 2001, Pages 545-559.
5. F. Mervyn, A. Senthil Kumar, S.H. Bok, A.Y.C Nee: "Developing distributed applications for integrated product and process design", Computer-Aided Design, Volume 36, Issue 8, July 2004, Pages 679-689.

6. Barone S., Beghini M., Bertini L. "Coupled Structure Analysis and Geometric Modelling in Rule-based Shape Optimisation". Proc. of The 11th ADM International Conference on Design Tools And Methods in Industrial Engineering, Vol. A., pp. 169-176, 1999
7. Rosen D.W., Grosse I.R., "A feature based shape optimization technique for the configuration and parametric design of flat plates". Engineering with Computers, Vol. 8, pp. 81-91, 1992
8. Takezawa A. et al. "Structural optimization for concurrent design and analysis at the conceptual design stage". In Advanced Design, Production and Management Systems, A.A. Balkema Publishers, pp. 9-16, 2003
9. Spath D., Neithardt W., Bangert C. "Integration of Topology and Shape Optimization in the Design Process". Proc. of the 2001 International CIRP Design Seminar, 2001
10. Bordegoni M., Cascini G., Filippi U. and Mandorli F., "A methodology for evaluating the adoption of knowledge and innovation management tools in a product development process", ASME Design Technical Conferences, Chicago September 2-6 2003.
11. G. Cascini, P. Rissone, F. Rotini: "From design optimization systems to geometrical contradictions", accepted for publication into the Proceedings of the 7th ETRIA TRIZ Future Conference, Frankfurt, Germany, 6-8 November 2007.
12. Cascini G., Rissone P.: "Plastics design: integrating TRIZ creativity and semantic knowledge portals", Journal of Engineering Design, vol. 15, no. 4, August 2004, Special Issue: "Knowledge Engineering & Management Issues in Engineering Design Practices", pp. 405-424.

# Comparison of Strategies for the Optimization/Innovation of Crankshaft Balance

Albert Albers[1], Noel Leon[2], Humberto Aguayo[2] and Thomas Maier[1]
1 IPEK – Institute of Product Development Karlsruhe, University of Karlsruhe, Germany
2 CIDT, – Center for Innovation in Design & Technology, Tecnológico de Monterrey, Mexico

**Abstract.** Engine crankshafts are required to be balanced. The balance of a crankshaft is one of several parameters to be analyzed during the design of an engine, but certainly a poor balance leads to a low life time of the whole system. It is possible to optimize the balance of a crankshaft using CAD and CAE software, thanks to the new optimization tools based on Genetic Algorithms (GA) and tools for the integration of the CAD-CAE software. GAs have been used in various applications, one of which is the optimization of geometric shapes, a relatively recent area with high research potential. This paper describes a general strategy to optimize the balance of a crankshaft. A comparison is made among different tools used for the sustaining of this strategy. This paper is an extension of a previous paper by the authors [1] but now different tools are being included to improve the performance of the strategy. The analyzed crankshaft is modeled in commercial 3D parametric software. A Java interface included in the CAD software is used for evaluating the fitness function (the balance). Two GAs from different sources and platforms are used and then they are compared and discussed.

## 1 Introduction

This paper compares different optimization and integration tools that can be used to build up a design strategy on crankshaft design and optimization. Though advanced CAD software has its own optimization engines, these can not be compared to more powerful genetic algorithms, such as those inside DAKOTA (Design Analysis Kit for Optimization Applications) [2] developed at Sandia Laboratories or even VGGA (Virtual Gene Genetic Algorithm)[3] developed at Tecnológico de Monterrey. It is therefore necessary to develop a proper interface to link the GAs to the CAD models. This work compares various approaches to this task, from commercial software to a JAVA interface required to integrate the genetic

---

*Please use the following format when citing this chapter:*

Albers, A., Leon, N., Aguayo, H., Maier, T., 2007, in IFIP International Federation for Information Processing, Volume 250, Trends in Computer Aided Innovation, ed. León-Rovira, N., (Boston: Springer), pp. 201-210.

algorithms with the CAD model of the crankshafts. A comparison of recently obtained results is presented in this paper.

### 1.1 Known works

The splinization approach to optimizing designs using GAs is relatively new; some examples are the computer design and optimization of cam shapes for diesel engines [4]. In this case the objective of the cam design was to minimize the vibrations of the system and to make smooth changes to the splined profile. Another example is the work developed at the Institute of Product Development (IPEK) at the University of Karlsruhe on the search for an assembly array of several tubular components whose trajectory is described by splines.

### 1.2 Our work

This paper continues from a previous work: "Computer Aided Innovation of Crankshafts using Genetic Algorithms" [1], where a method for determining the design imbalance of CAD modeled crankshafts inside CAD software is described. In this previous work a procedure for a balanced design strategy was introduced. The mass properties required to calculate the balance ($m_g$, $r_{gy}$, $r_{gz}$ and the inertia products $I_{xy}$ and $I_{xz}$) of the crankshaft models can be obtained from parametric CAD software, which has special commands in its advanced modules, for calculating the balance as a response for fitness evaluation. If a target for imbalance is specified the difference between the target and the current imbalance is a fitness function that has to be minimized by making modifications to the crankshaft geometry. In order to make geometry modifications it was decided to substitute the profile of the counterweights with splines from its original "arc-shaped" design, because cubic splines allow smooth shape changes via control points with continuous second derivatives, a desired property for material fluency during the manufacturing process. The x and y coordinates of the control points can be parametrically manipulated by an optimization algorithm, i.e. by a Genetic Algorithms (GA).

## 2 Preliminaries

### 2.1 Problem Description

The figure below shows a counterweight profile of a V6 engine crankshaft. From a previous sensitivity analysis it was found that defects in the profile (under filling) of the outermost counterweights, number 1 (CW1) and 9 (CW9), have the greatest influence on the balance of the crankshaft. It was decided to begin with codification of the "Y" coordinates of 4 control points, each upper and lower profile from both counterweights, resulting in 16 control parameters. The variation of these control parameters results in a balance response. The fitness function selected as the response is an equally weighted function of the differences between the specified target for imbalance and the current imbalance on the two external sides of the crankshaft (CW1 and CW9).

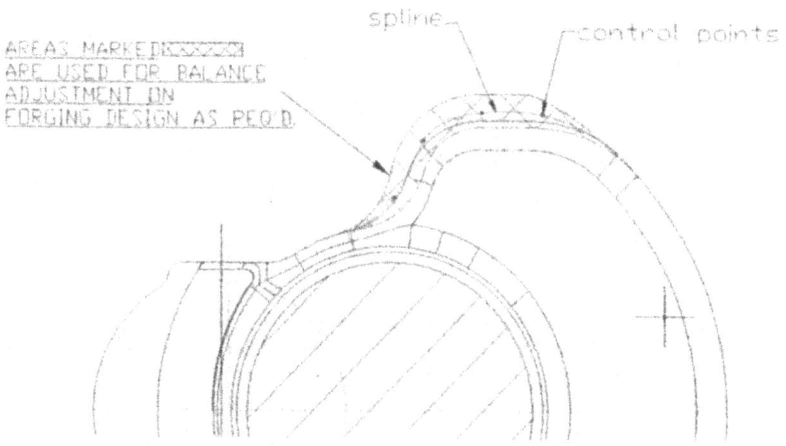


**Fig. 1.** Profile of a counterweight represented by a spline and its control points

This method is called "Goal Programming", in which the designer has to assign targets or goals to be achieved for each objective function [5]. These values are incorporated into the problem as additional constraints. The algorithm will then try to minimize the absolute deviations from the targets to the objective functions. The simplest form of this method may be formulated as follows:

$$\min \sum_{i=1}^{k} |f_i(x)-T_i|, \text{ subject to } x \in X$$

where $T_i$ denotes the target or goal set by the designer for the ith objective function $f_i(x)$ and X represents the feasible region. The criterion, then, is to minimize the sum of the absolute values of the differences between the target values and the actually achieved values of imbalance on the two external counterweights.

### 2.2 Strategy and Approach No. 1: CAD Model linked to VGGA via Commercial Integration Software

Using VGGA, a population of 50 individuals is generated randomly. Each individual is a gene containing a set of 16 parameters, codified in lineal mapping 0s being the lower limits and 1s the upper limits. Once generated, the phenotypes (decodified) are taken by the commercial software, parsed to the macro file read by the CAD software and run automatically to generate a fitness result; this fitness result is read and sent back to VGGA, where it is assigned to the corresponding individual.

After the whole generation has been evaluated, tournament selection plus other genetic operators produce the next generation. Figure 2 shows the proposed sequence of steps.

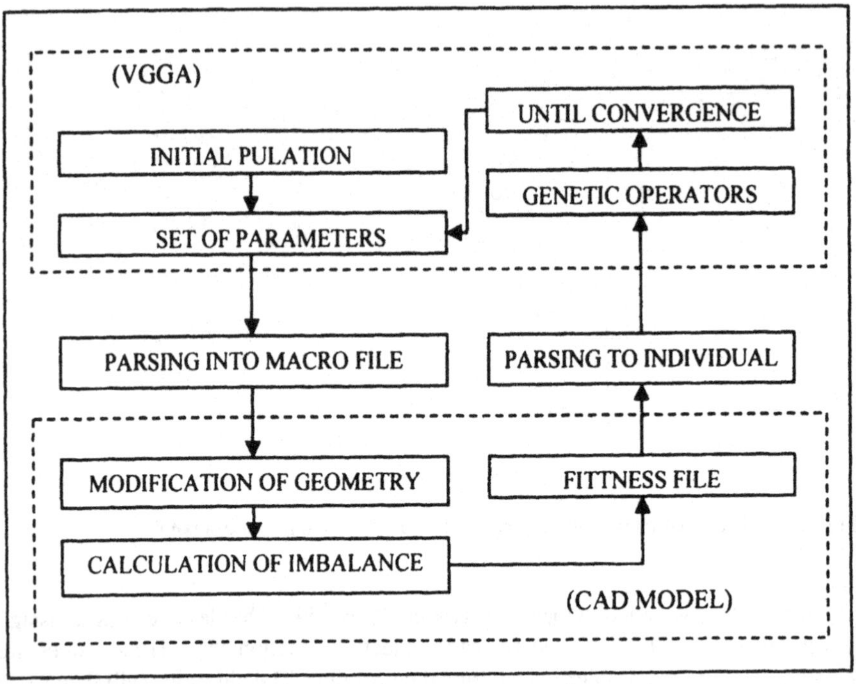


**Fig. 2.** Integration of VGGA with the CAD Model using commercial software

The Genetic Algorithm performed well during the initial runs, as it was capable of delivering a geometry that improved the fitness function in each generation. However, because commercial software was used for the integration, there was low flexibility and the CAD software started up on each individual evaluation, causing a long standing time, which made large runs prohibitive. Two additional approaches for the same strategy are presented, and a comparison has been made.

## 3 Development of Alternative Approaches

### 3.1 Strategy and Approach No. 2: CAD Model linked to VGGA via JAVA

The parametric CAD software used in the project comes with an interface programmed in JAVA, called J-Link, and some of the subroutines were adapted to substitute the commercial software from the first approach, in a sequence similar to that shown in figure 2 but with some differences:

1. The control of the Genetic Algorithm is still made by VGGA, but the evaluation of the fitness function in the CAD software is controlled by JAVA, instead of the commercial integration software.
2. The CAD software is kept running in parallel to VGGA. The CAD software therefore does not need to stop its execution and start up every time the fitness function is evaluated. Once the fitness function file is ready, VGGA reads it and assigns it to the corresponding individual to continue with the algorithm.

The main advantage of this approach was that the need to start up the CAD software after each evaluation is eliminated, and it shut downs and starts again automatically only when the set of parameters causes a geometry that cannot be regenerated (for example, a CAD model that conflicts with characteristics form areas other than counterweights). When this happens, the fitness function is assigned a value of zero so that the individual does not to continue to the next generation. Some runs were made with a similar set of parameters as with the commercial software. Being able to run a substantial number of evaluations, VGGA showed a convergence to a fixed fitness value, as shown in figure 3.

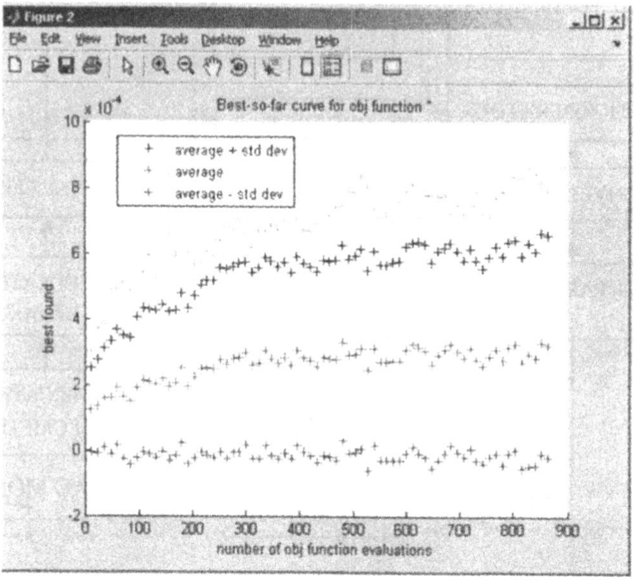


**Fig. 3.** Evolution of the fitness function using VGGA

It should be noted that the graph does not show a continuous increase, which would have been expected as the balance reaches the target. Because the target balance could not be attained inside the design constrains defined for the geometry of the counterweights, some trials were performed allowing the spline to trespass the geometry constraints. However, although improved results were obtained, the convergence was not yet satisfactory. It should be noted that the shapes that trespass the boundaries, although not satisfactory from the forging point of view, tend to "separate" the balancing mass out from the crankshaft center axis. This kind of solution has already been implemented in the design of other engine crankshafts. This behavior leads to the conclusion that a different algorithm was required for validating the convergence. It was therefore decided to substitute VGGA, and to continue to the next step of the study using a GA solver inside the DAKOTA toolkit.

## 3.2 Strategy and Approach No. 3: CAD Model linked to DAKOTA via JAVA

DAKOTA is a parameter solver developed at SANDIA Laboratories with the original goal of providing a common set of optimization tools for engineers who need to solve structural and design problems, including Genetic Algorithms. The proposed sequence of steps is shown in figure 4, which shows the parallel execution of DAKOTA and CAD Software via JAVA.

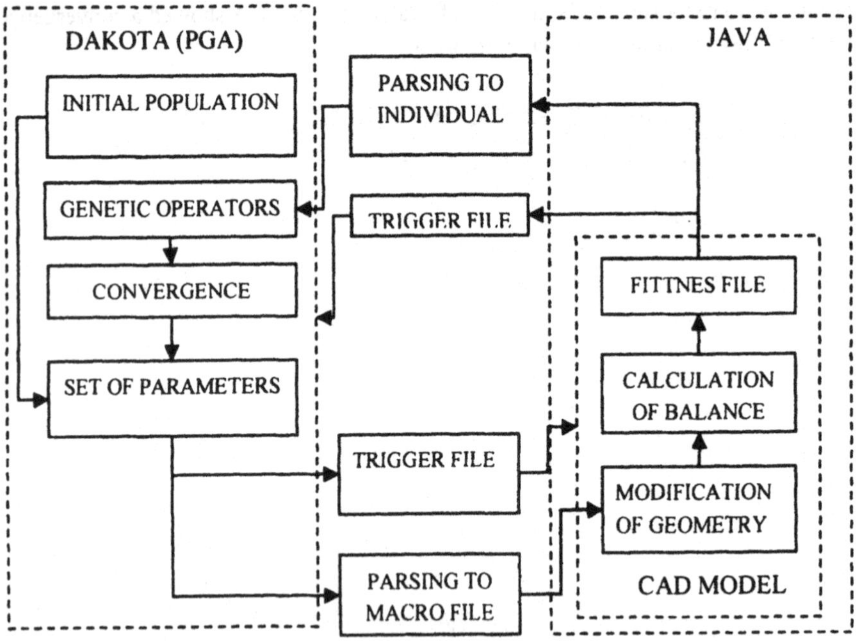


**Fig. 4.** Integration of DAKOTA with CAD Software using JAVA

In order to substitute VGGA by a DAKOTA GA it was necessary to make some adjustments to the strategy:

1. A DAKOTA GA was chosen that solves for minimization, instead of making a maximization, so the fitness function is the inverse of the result used in VGGA, aiming at a zero value (important when comparing the two evolution graphs).
2. Most of the parameters used in VGGA were kept (mutation and crossover ratios, etc) but now the solver (named pga_real) is from the DAKOTA

"Stochastic Global Optimization" library, a genetic algorithm for real numbers.

The evolution of the fitness function converged to a non-zero value, as can be seen in figure 5. The formatting of the graph is different than in VGGA, because different graph-making tools were used.

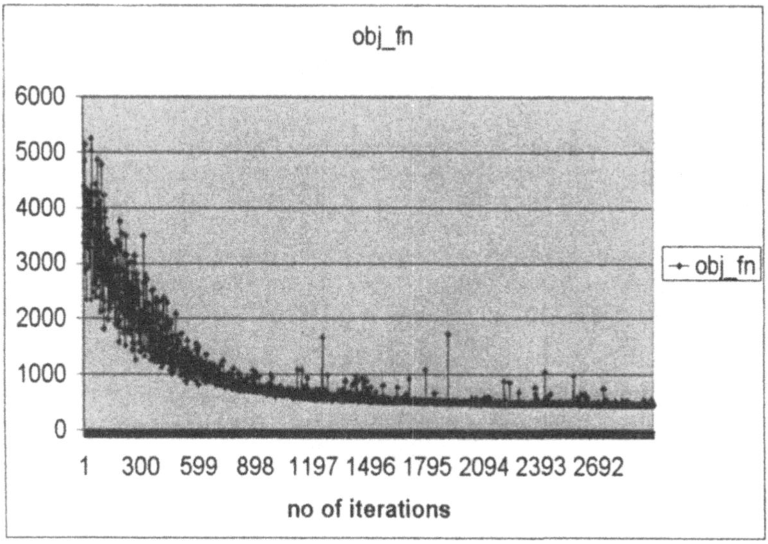


**Fig. 5.** Evolution of the fitness function using DAKOTA

It should be noted that it was not possible either with VGGA or with DAKOTA, to reach the balance target inside the design constrains of the parameters that control the splinized counterweight profiles. After analyzing the balancing behavior of some individuals during the evolutions of the two GAs, it was observed that when a good balance was reached in CW1, the balance in CW9 worsened, and vice versa. This conflicting behavior needed to be confirmed using the Pareto Frontier method.

The balances on both sides of the crankshaft (CW1 and CW9) were defined as two independent objective functions. Because no data on this condition had been collected on the previous runs, a new run was required. We chose a Multi Objective Genetic Algorithm (MOGA) from the DAKOTA toolkit to generate the required data. Figure 6 shows a graph with the value of the first function (imbalance of CW1) in the x-axis and the value of the second function (imbalance of CW9) in the y-axis. This graph is known as the Pareto Frontier

There is a generalized notion in multi-objective optimization that no "optimal" solution can be attained but that there is a set of optimal solutions lying on a line that prevent the functions reaching the "ideal" at the same time. Here, different theories on "trade off" management arise. Further discussion can be conducted on how innovation theories can "jump over" the technical contradiction and how Genetic Algorithms could help in this issue. From the point of view of a classical genetic algorithm the closest approach dealing with the "ideal" concept is related to an

"Alternative Ideal": to find the "ideal point", a decision must be taken by exploring the limits along each constraint so that an alternative ideal can be defined [5].

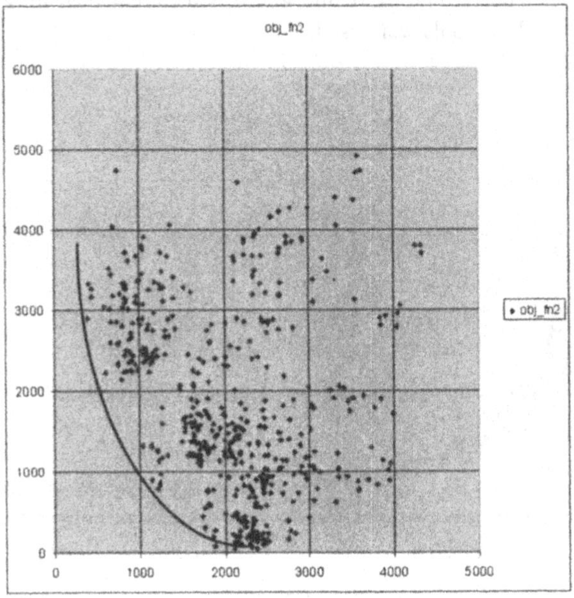


**Fig. 6.** Pareto Frontier between conflicting functions

In order to have a visual feeling about the way the algorithm is performing, an individual from the most centered value on the Pareto Line of the graph was selected. The resulting geometry of that individual is shown in figure 7. In these pictures the profiles of the counterweights come close to the limits imposed by the original design (sketched in blue). It is inferred that, in order to reach the balance target, it may be necessary to reconsider the geometric constraints. As can be seen in this case, sharp edges arise in the profile, and this is not good for the forging of the crankshaft. New constrains to the optimization strategy have to be formulated, taking control of the curvature of the splines in such a way that no sharp edge develop and forging process is not worsened. This is only a single step forward in a field of specific functionality, but expanding the solution to open design dimensions [6], there will be a need for factors that exploit the parametric geometry of the CAD system with innovative concepts.

The proposed system has two objective functions (imbalance-to-target on CW1 and imbalance-to-target on CW9), and the original geometric constraints are widened (lower and upper limits of the control points of the splines, plus constraints on the curvature of the splines.

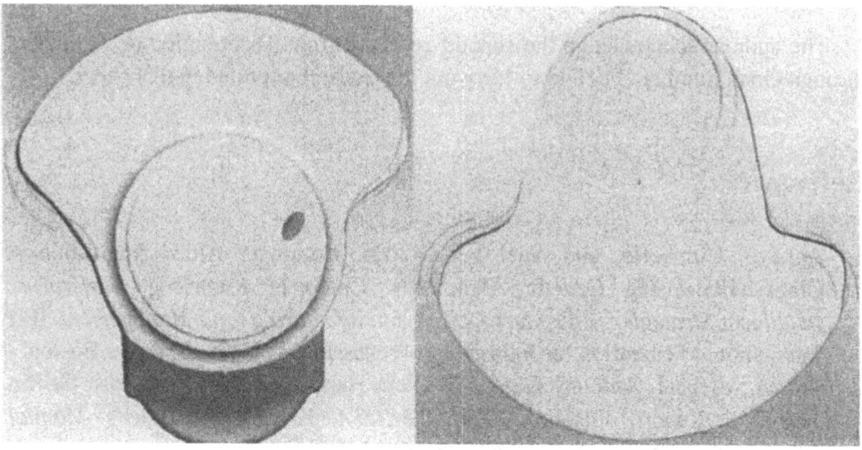

**Fig. 7.** Profile Shapes of CW1 and CW9 from an individual in the Pareto Frontier

## 3 Conclusions and further research

Both VGGA and DAKOTA converged in about the same number of evaluations. The use of the JAVA interface prevented the need to start up the CAD software during each evaluation that significantly improved the run time. MOGA from DAKOTA allowed developing the development of a Pareto frontier from which important conclusions were extracted:

- It is necessary to prevent the sharp edges of the splines by adding an extra restriction that controls the curvature of the shape. This can be done by adding an evaluation during each individual run, in an additional solver.
- Simulation of the forging process is a required next step in order to define a relationship between good shapes-curvature and manufacturability. This becomes significantly important when a proposed design outside the initial shape restrictions needs to be justified in order not to affect formability.
- Stress analysis of the crankshaft is a preferred approach to be added either a restriction or an objective during modification-evolution of the geometry of the crankshaft in order not to affect its functionality.

This paper defines the basis and the beginning of a strategy for developing crankshafts that will include the manufacturability and functionality to compile a whole Multi-objective System Optimization. The use of the Java interface will allow the control and manipulation of additional software to be required during restriction evaluations. A medium-term objective is to implement innovation concepts, i.e. operators, into the evolutionary mechanisms of computer algorithms in order to trespass the barriers imposed by conflicting objectives (Pareto frontier).

## Acknowledgements

The authors acknowledge the support received from Tecnológico de Monterrey through Grant number CAT043 to carry out the research reported in this paper.

## References

1. Aguayo, Humberto, and Noel Leon. 2006. Computer Aided Innovation of Crankshafts Using Genetic Algorithms. Chap. in *Knowledge Enterprise: Intelligent Strategies in Product Design, Manufacturing, and Management*. IFIP International Federation for Information Processing. 471-476. Springer Boston.
2. Eldred, Michael, Anthony Guinta, William Hart, John Eddy, and Josh Griffin, Technical Report. October 2006. *DAKOTA Version 4.0 User's Manual.* Albuquerque, NM, USA: Sandia National Laboratories.
3. Valenzuela-Rendón, Manuel. The Virtual Gene Genetic Algorithm. Chap. in *Genetic and Evolutionary Computation GECCO 2003.*
4. Lampien, J. JULY 2003. Cam shape optimisation by genetic algorithm. *Computer-Aided Design* 35, 8: 727-737.
5. Coello Coello, Carlos Artemio. 1996. An Empirical Study of Evolutionary Techniques for Multiobjective Optimization in Engineering Design. Doctor of Philosophy diss., Tulane University.
6. Cueva de Leon, Jose Maria. 2006. Automatic shape variations for optimization purposes. Master in Sciences thesis, Technologico de Monterrey (ITESM).

# Collaboration in automotive winter testing

*Real-time simulations boosting innovation opportunities*

Mikael Nybacka, Tobias Larsson, Åsa Ericson
Division of Functional Product Development, Luleå University of Technology

**Abstract.** The performance of cars has during recent years become increasingly dependent on complex electronic systems used especially for safety but also comfort, performance and informatics. Automotive winter testing activities in northern Sweden is vital to test and try out those systems. A contradiction to increased performance is that faulty software also causes 30 % of severe malfunctions in the functionality of the car. To deal with these problems, as early in the design process as possible, innovative methods to cope with digital abstraction and the physical world in a unified way seems promising. One useful approach, in automotive winter testing, might be to support the possibilities for real-time vehicle simulations of the car in motion.

The closer collaboration in the automotive industry might be an incitement for investing in technologies for knowledge sharing. Besides enhancing the product development process, additional knowledge might support innovations. Today, instead of providing parts similar to their competitors and relying on one or two automakers, successful suppliers focus heavily on innovation and on collaboration with a number of manufacturers on a global market. Due to the possibilities to visualize whole processes, the use of simulations seems to support a 'seeing first' approach to innovations.

Thus, the purpose of this paper is to describe an as-is scenario and a to-be scenario for automotive winter testing to highlight how the use of real-time simulations facilitates innovative methods.

## 1 Introduction

In the northern part of Sweden the winter season starts in November and lasts into March, temperatures in the range of - 40 degrees Celsius are common in some lower areas 1. The natural asset - cold weather - is considered as a possibility and has become a backbone in many companies' business concepts and particular in services for winter car testing. The four municipals, Jokkmokk, Arvidsjaur, Arjeplog and Älvsbyn, constitute a vast region where collaboration is important to support the automotive test industry 2. The collaborative effort takes place in terms of providing

*Please use the following format when citing this chapter:*

Nybacka, M., Larsson, T., Ericson, Å. 2007, in IFIP International Federation for Information Processing, Volume 250, Trends in Computer Aided Innovation, ed. León-Rovira, N., (Boston: Springer), pp. 211-220.

overall services surrounding the test industry, in terms of for example logistics, accommodation, food and leisure activities. Within this region, a number of automotive winter test service companies are established.

The area is sparsely populated, but this fact is also turned into a potential to provide added value for the testing industry. Besides many kilometers of a wide variety of low traffic public roads and ice tracks on natural lakes, huge proving grounds especially designed for automotive winter testing are provided for testing of cars and components 13,4.

In this setting, the automotive test entrepreneurs' services are vital in the collaborative efforts in the design and development of cars, e.g., automotive manufacturer and suppliers. Automotive manufacturers, i.e., Original Equipment Manufacturer (OEM), are for instance, General Motors and Fiat 3. Suppliers, in this case, are usually those who provide main parts, i.e., Tier 1 suppliers, meaning that they, in turn, purchase components from small part makers. Examples of main part providers are Haldex 3, Knorr-Bremse 4, TRW Automotive 5, Bosch and BMW with own test facilities. Automotive manufacturers in the United States are reducing the number of suppliers to compensate for lower market share. For example, Ford is halving the number of suppliers from which they buy seats and wiring 6. Furthermore, Ford has identified 43 suppliers for close collaboration, 13 of those have been contracted on a long-term basis or where contracts already exists these has been extended over many years 5. This closer collaboration might be an incitement for investing in technologies for knowledge sharing. Besides enhancing the product development process, additional knowledge might support innovations. Today, instead of providing parts similar to their competitors and relying on one or two automakers, successful suppliers focus heavily on innovation and on collaboration with a number of manufacturers on a global market, 6. Due to the possibilities to visualize whole processes, the use of simulations seems to support a 'seeing first' approach 8 to innovations.

The performance of cars has during recent years become increasingly dependent on complex electronic systems used especially for safety but also comfort, performance and informatics. Automotive winter testing activities in northern Sweden is vital to test and try out those systems. A contradiction to increased performance is that faulty software also causes 30 % of severe malfunctions in the functionality of the car 9. To deal with these problems, as early in the design process as possible, innovative methods to cope with digital abstraction and the physical world in a unified way seems promising. One useful approach, in automotive winter testing, might be to support the possibilities for real-time vehicle simulations of the car in motion.

Thus, the purpose of this paper is to describe an as-is scenario and a to-be scenario for automotive winter testing to highlight how the use of real-time simulations facilitates innovative methods.

### 1.1 Functional Product Innovation – a framework

A framework for the plausible to-be scenario is found within a Functional Product Innovation (FPI) vision. The vision is a joint academic and industrial construct to capture a widening in view among manufacturing companies, the view widens from focusing mainly on the physical artifact to also entail a view on product development where the performance of the physical product is provided as a service. The goal is to take cross-company knowledge domains of engineering, business and production into account in the design phases. This vision puts an emphasis on additional knowledge and information in early design phases, for example understanding of the actual use of the product and the environment where it is going to be used is important, since these aspects needs to be designed into the final product. Life-cycle perspective and close cross-company collaboration in the design and development of products constitutes a basis for realization of FPI. The collaborative efforts and widening in view are thought of as a facilitator for innovations to arise. Furthermore, a simulation-driven approach in early phases to support decisions in product development, by the same token, try out those solutions in numerous of what-if business scenarios, is also included in the vision. Mintzberg and Westley 8 write; *"...As Mozart said, the best part of creating a symphony was being able to see the whole of it in a single glance in my mind"* (p.90). A 'seeing first' approach 8 is vital in an innovation process and visualization is an underpinning idea in presentation of simulation result. As a complement to validation, the early use of simulations is thought of as supporting a virtual structure to combine and recombine knowledge from the collaborative partners.

### 1.2 Innovations - a wider perspective

An understanding, we are not attempting for a definition, of the word innovation seems a useful starting position for this paper. The word innovation is used oftentimes as new physical artifacts or commodities, i.e., *new things*, which has reached a market and satisfy some sort of needs. In a FPI context, the extended product definition adds *new services* to the word innovation. In turn, in the service performance, the users of the service act as co-producers in the development process. Hence, *new processes* and *new methods* to carry out design activities arise and can be included in the word innovation. Such a collaborative product development brings in the qualities *new knowledge* and *new ideas* into the understanding of the word innovation.

The word *new* can here be interpreted as in beforehand 'poorly understood' or 'unknown' and as a fact, exceeding what was intended from the beginning, thereby understood as innovation. The frame for discussing innovations in this paper is delimited to new methods such as new methods for performing automotive winter testing.

## 2 Method

In general, empirical data for the study presented in this paper has been generated during informal and formal meetings with companies in the automotive winter testing industry. People from OEM, Tier 1 suppliers and service entrepreneurs have been involved in these meetings. The form of data generated by participation in these meeting is mainly qualitative, e.g., an interpretation of something in the context where it occurs 10. Due to the participative approach, this study might be described as related to action research 11.

A survey, for generation of both quantitative and qualitative data, has been performed in addition to this, also including all three levels of actors. The qualitative part result of this survey is used in this study to build the as-is scenario and to discuss the to-be scenario, as well as a complement to the findings in the meetings. The results from the survey as a whole will be presented in a forth-coming paper.

The technological set up for simulation to support collaboration is built on the work of Larsson et al 12, Larsson and Larsson 13 and Törlind 14, and an initial technical description for real-time simulations can be found in Nybacka et al 15, 16.

## 3 Automotive winter testing facilities – views from the actors

Compared to their competitors, entrepreneurs in the northern Sweden test region highlight *"relatively stable winter"* and *"big areas on lakes for ice tracks"* as important advantages to provide and enhance the test services. Moreover, entrepreneurs think that the ability of the locals to speak not only English, but also German can be considered as an advantage for the test region as such.

The entrepreneurs say that the fact that large Tier 1 suppliers are established and located in the area is an important contribution to the competitiveness, since they attract OEMs to accommodate their automotive winter testing activities in the region. One entrepreneur says that *"firmly established big Tier 1 suppliers function as a magnet for the OEMs"*. The entrepreneurs perceive the transportations of cars from other parts of the European Union as easy, but they also consider the poor travel possibilities for people (flights and long distance to the testing area) as a negative aspect. The entrepreneurs bring together sub-contractors who are directly involved with working to keep the test tracks, etc. in good condition, along with additional firms that provide added value to leisure time, e.g. firms that provide snowmobile safaris or dog sleighing tours. The entrepreneurs perceive a risk in that the increased leisure events take capacity from the test activities.

In a future perspective of five years, the entrepreneurs believe that they will probably sell more services and that the OEMs will join and share test facilities. Furthermore, they hope that there will be summer testing and that the test activities will continue to positively develop.

For automotive winter testing in the northern Sweden, Tier 1 suppliers express the advantages as having OEMs together. *"This enables us to have demos with significant number of participants and it also helps collaboration in development within customer projects"*. One of the respondents expresses that the entrepreneurs offer *"perfect service"* and that they *"know what the customer needs"* and provide well-prepared facilities. Another respondent emphasizes the climate as an advantage for winter tests, *"Northern Sweden has a climate that is very suitable to winter testing. It is guaranteed to get cold, new snow and big temperature variations if you stay approximately seven to ten days"*. This is a reason why this respondent prefers the northern Sweden test facilities in favor of competitors in Finland and Canada.

However, the OEMs and Tier 1 suppliers highlight that traveling to the test sites is time consuming, and that it is difficult to get hotel rooms, especially with short notice.

OEMs and Tier 1 suppliers see changes in the test processes as two-fold. Firstly, they forecast a reduction in the number of products, but an increase in the total amount of testing hours. This seems likely because, as they express, *"more time will be spent on each product that should be tested"*. Secondly, they see a general reduction of testing activities, since tests can also be performed in cold climate chambers. However, they find the first trend as plausible because, as one of the respondents reflects on changed processes, *"I believe that the tests will be standardized with well-defined methods, which in turn will reduce the number of people needed for the test. With good methods for performing winter testing, I also believe that the tests could be outsourced to the entrepreneur. This will reduce the costs"*. He continues that if the entrepreneur is hired to perform the tests, the car can run *"life length test for a longer period and in turn reduce costs and get more value from the test season"*.

## 4 A setting for a component test – a typical case

The OEM provides the cars where the test equipment is mounted by the Tier 1 supplier. The Tier 1 supplier is responsible for the testing as such. When a test is accomplished, the OEM receives a test report from the supplier's head office.

Typically, the component test involves groups of people performing the test assignment. The Tier 1 supplier testing activities take place on site, e.g., on the proving ground in the North of Sweden in collaboration with off site staff, e.g., the product development head quarters located outside Sweden. On site staff, are for example:

- Test drivers possessing expertise about specific components.
- Team leader who coordinates the test sessions and sends or transfer test data to home office.
- Mechanics that repair or mount test equipment and extract data from test vehicle.

Tier 1 supplier off site staff are for example, managers and system experts related to the electronic control system, but also engineers and designers.

The service entrepreneur operates locally on site being responsible for providing a purposeful proving ground, e.g., communication, different kinds of tracks, preparation of tracks, gas stations, garages, car washes, cold chambers and transportation within the area. They are also providing information about e.g., track surface conditions (ice tracks on natural lakes changes constantly due to weather conditions), weather forecasts, and booking services for travel and accommodation.

One or more cars are driven at the test facility and different types of data are gathered depending on what is under scrutiny, for example data about steering, speed and acceleration. The test driver usually change parameters during a test run, however, if any larger corrections are needed, the test driver has to pull up to the garage and stop the car so that the team can make those corrections.

## 5 As-is scenario – current test procedure

Different types of sensors record and sample data from the vehicles. The logging of data is done independently in each vehicle, and is gathered from the vehicles respectively after the test drive. Thereafter, system experts put together and analyze the data, which are visualized in tabled or graph form. In general, the test process is conducted in steps; test driving, analyzing the data and decision-making, see Figure 1. Decision about whether or not a new test is needed, is based on what is found. The conditions for that new test have to be in resemblance to the previous test. Similar conditions are considered as hard to reproduce. The conditions can be according to one aspect or a combination of aspects, for example outside temperature and/or the ice surface topology. Finally, the system experts write a report presenting data and findings.

Several problems can occur during the tests, for instance failing sensors, control systems and/or human errors. These problems can be hard to notice by the test driver even in those cases when they have a laptop computer on the passenger side displaying test data. The test driver has to focus on driving the car and cannot simultaneous analyze what is displayed. Therefore, despite having a laptop, the test drivers may not receive any direct feedback of the recording process when driving the car; hence, failing sensors or systems are discovered after the drive. Due to a lack of feedback to the test drivers, they do not have any possibilities to directly stop or change the test drive, the control system or analyze the situation.

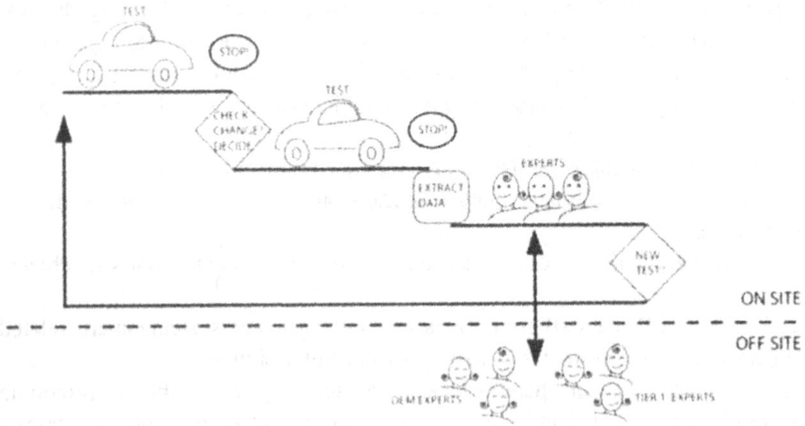


**Fig. 1.** An example of as-is procedure.

# 6 To-be scenario – discussing a future test procedure

The way automotive winter testing are performed, a test driver sitting in the car driving on tracks and experts sitting outside the car, even though on site, are by its nature distributed, i.e., there is a physical distance between the actors. The basic idea for a future component test scenario is to create a real-time visualization application to support the distributed work technically. By doing so, off site staff at the home office as well as other experts from locations all over the world can be included into the test sessions in real-time. Since divergent knowledge areas, as well as organizations (OEM, Tier 1 and entrepreneur) need to collaborate, the communication can be viewed as crossing boundaries. Visualization, or a 'seeing first' approach to decision-making is supportive when the situation is new 8. The distributed real-time visualization application seems to be a way to support collaborative decision-making and enhance the concurrency between activities.

The structure for this suggestion for how to support automotive winter testing activities is web-based. In Figure 2, an overview for the suggestion is shown. Data from the test vehicle are transmitted by a wireless access, recorded and stored at a server from which data are retrieved for analyses. The ability to see the result of the test in 3D can be provided by the use of advanced 3D engines, such as AgentFX™ 17. A unique advantage of AgentFX™ is that it allows complete control over the rendering process, allowing creation of advanced user interface, with 2D overlays and augmented 3D views. This capability can support a new test to be set up in a similar way as a previous one.

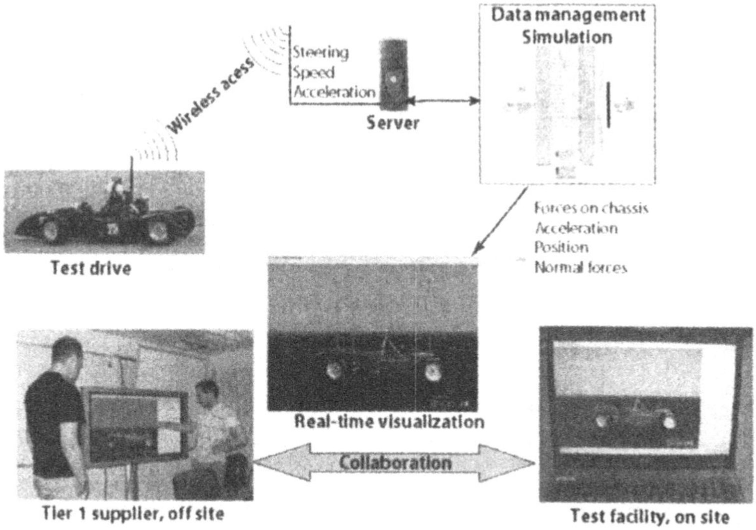


**Fig. 2.** A structure overview.

The technical structure and the real-time distribution of measurement data have been tried out during a winter testing session at Arctic Falls test track 1. The measurement data were temperature data from the car that were visualized in a virtual environment 14. The framework with dynamic simulation software, Matlab/Simulink and Java based visualization application viewed in figure 2 have been and are currently further tested in laboratory settings 1516. In addition to this, cooperation with another research project [18] resulted in a successful test of saving streamed visualization, video and audio data synchronized.

In a to-be scenario, the test driver log-on via a laptop to a secured website where the visualization application is hosted. During the test, all staff, on site and off site, can in a 3D environment follow the test; this direct feedback verifies that the test is being logged and analyzes of the test data can start concurrently in a more direct and collaborative way. Moreover, the test staff can directly discover extreme system behavior and potential errors. Besides representation of data in 3D, regular data presented in graphs or tables can be used. On the basis of these analyses, the system experts can decide if a retest is required, thus more easily done during similar conditions. The system experts can via a video or audio link to the test driver give instructions for e.g., retest, where 2D overlays and augmented 3D views support setting up and performing the test under similar conditions as a previous test.

Due to the direct connection from the vehicle to the server, data can be stored in a structured way and also be replayed after the test session. Besides making it possible to further analyze a noted problem, the recently logged data can be compared with past data.

In this way, the test procedure has changed from being sequentially performed to being performed concurrently with the analyses and decision-making processes, see Figure 3.

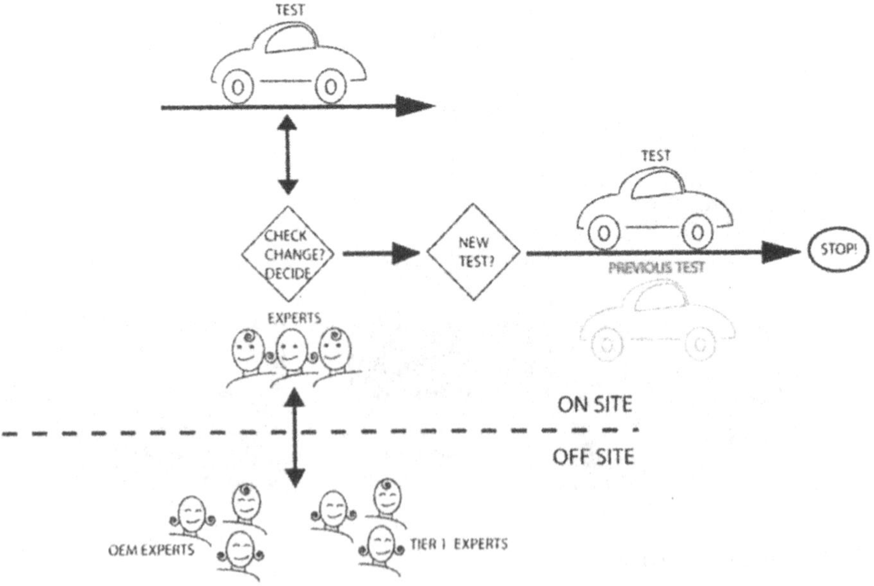


**Fig. 3.** An example of a to-be procedure.

This distributed technical solution supports a new test procedure. Besides, probably, bringing with it innovations in terms of both software and hardware products, it opens up for new services which the entrepreneurs might provide. One respondent, from the Tier 1 supplier and OEM group, have expressed a change in testing activities as becoming standardized and with well defined methods. He sees that the number of people needed for testing will be reduced and believes that the entrepreneur could run the cars for life length test for a longer period. In this way, the entrepreneurs become more involved in the actual vehicle tests, and this might give possibilities to build up technological capital, i.e., a base for new services, and to extend the activities at the test site during the rest of the seasons.

The respondents have expressed travels as a bottleneck, because of taking too long time. A distributed solution enables people to collaborate, but travel less.

## 7 Concluding remark

In this paper, an as-is scenario has been described and a to-be scenario of automotive winter testing has been presented and discussed for the application of real-time simulations as a facilitator for innovative work methods during automotive winter testing. By supporting the to-be test procedure with real-time simulations and 3D visualization in particular, the way of conducting the tests radically changes to a more concurrent test process. Also, distributed collaborative work is made doable. In addition, the approach enables decision-making to become a more concurrent activity since off-site experts can take part simultaneous with the testing activities. The approach raises opportunities to extract rich information of vehicle and its systems, which provides a good basis for well informed decisions. The possibility to save streams of visualization, video and audio data synchronized from the vehicle facilitates gathering of drivers' actions and feelings which can be linked to the behavior of the vehicle displayed in the 3D environment. The connection between the vehicle running the test and the development office will be a good base for new future innovation concerning vehicle validation and development. In turn, your car will be safer to drive.

Besides continuing evaluation and verification of this research, further considerations on how to apply the principle of distributed work supported by real-time simulations in other areas are interesting within the FPI vision. In the light of a simulation-driven approach, where simulations are thought of as supporting a virtual structure to integrate divergent knowledge areas, e.g., business, engineering, production, in order to drive product development - how can results from real-time simulations drive the design of next generation products and can it be used to enable new business concepts?

## Acknowledgments

Fundings from Center for Automotive System Technologies and Testing through Norrbottens Forskningsråd, as well as the Faste Laboratory, through VINNOVA and partner companies are greatly acknowledged.

## References

1. Arctic Falls, (March 15, 2007); http://www.articfalls.se.
2. Swedish Proving Ground Association, (March 15, 2007); http://www.spga.se.
3. Colmis, (March 15, 2007); http://www.colmis.com.
4. Car Test, (March 15, 2007); http://www.cartest.se.
5. Cold Climate Technology, (April 20, 2007); http://www.wintertest.com/wintertest/index.htm.
6. N. Bunkley, 2007. *Buyout shops see profit in distressed auto part sector*. Published 30 January 2007. Assessed 11 March 2007. International Herald Tribune. http://www.iht.com/articles/2007/01/30/business/place.php.
7. Ford Motor Company, (March 15, 2007); http://media.ford.com/article_display.cfm?article_id=25091.
8. H. Mintzberg, F. Westley, Decision making: it's not what you think. *MIT Sloan Management Review*, Spring 2001, 89-93.
9. J. Vincentelli in Shäuffele, T. Zurawka. *Automotive Software Engineering. Principles, Processes, Methods and tools*. (SAE International Warrendale, 2005).
10. J. Mason, *Qualitative researching*. (Sage Publications, London, 1996).
11. E. T. Stringer, *Action Research, 2nd edition,* (Sage Publications, USA, 1999).
12. T. Larsson, A. Larsson, L. Karlsson. Distributed Multibody Dynamic Analysis Within Product Development. In proceedings of the *DECT'01 ASME 2001 Design Engineering Technical Conferences & Computers and Information in Engineering*. September 9-12, 2001, Pittsburg, Pennsylvania.
13. T. Larsson, A. Larsson. Web-based Multibody Dynamics using Distributed Simulation Modules, Annals of 2002 *Int'l CIRP Design Seminar,* 16-18 May, 2002, Hong Kong.
14. P. Törlind. A Collaborative Framework for Distributed Winter Testing. In proceedings of *eChallenges e-2004*, October 2004, Vienna, Austria.
15. M. Nybacka, T. Larsson, M. Johanson, P. Törlind. Distributed Real-time Vehicle Validation. In proceedings of *IDETC/CIE 2006, ASME International Design Engineering Technical Conferences & Computers and Information in Engineering Conference*, September 10-13, 2006, Philadelphia, Pennsylvania, USA.
16. M. Nybacka, T. Karlsson, T. Larsson. Vehicle Validation Visualization. In proceedings of *Virtual Concept 2006*, November 26-December 1, 2006, Playa Del Carmen, Mexico.
17. Agency 9, (March 15, 2007); www.agency9.com.
18. Flexible collaboration tools for distributed engineering in automotive winter testing, (March 15, 2007); http://www.ltu.se/polopoly_fs/1.4090!75a8fda4.pdf.

# Future of CAE and Implication on Engineering Education

Seung-Hyun Yoo[1], Eung-Jun Park[2], Jae-Sil Lee[2], Joon-Ho Song[2], Dae-Jin Oh[2], Woong–Rak Chung[2], Yeongtae Lee[2] , and Dhaneshwar Mishra[2]
1 Department of Aerospace Engineering, University of Maryland, College Park, Maryland, USA, On leave from Department of Mechanical Engineering, Ajou University, Suwon, Korea ryseung@gmail.com
2 Department of Mechanical Engineering, Ajou University, Suwon, Korea
WWW home page: http://www.ajou.ac.kr/~sml

**Abstract.** In this paper, the trend of FEM which is one of core tools of CAE and the future of CAE are examined in terms of TRIZ concept. This observation leads to the necessity of engineering creativity. This paper reports our successful experiences of implementing systematic innovation tools into engineering design classes. Also potential revolutionary change of engineering education is hinted by extending the accomplishments of TRIZ. It can innovate the current engineering curricula and an on-going study of classifying the engineering activities is introduced.

## 1 Introduction

Every modern organizations needs innovation tools in order to adapt in rapidly changing environments, especially modern companies [1-3]. Radical innovations require three elements, namely, they should be new to market, 5~10 times more productive and show 30~50% cost reduction. It is generally accepted that current curricula of engineering education and engineering practices cannot meet these requirements. We have been trying to find a systematic tool for innovation, creative design, and inventions using CAE (Computer Aided Engineering) tools such as FEM (Finite Element Method) as the authors have been teaching structural mechanics courses in conjunction with engineering design. Current curricula of mechanical engineering consists of basic sciences such as mathematics, physics, chemistry, and biology (this course was added recently to expand current mechanical engineering areas) and mechanics courses such as statics, dynamics, solid and fluid mechanics, thermodynamics and design practices from machine elements design to system design. Those core disciplines are called engineering sciences and their primary goals are to understand the materials and systems in order to simulate and predict the

---

*Please use the following format when citing this chapter:*

Yoo, S.-H., Park, E.-J., Lee, J.-S., Song, J.-H., Oh, D.-J., Chung, W.-R., Lee, Y., Mishra, D., 2007, in IFIP International Federation for Information Processing, Volume 250, Trends in Computer Aided Innovation, ed. León-Rovira, N., (Boston: Springer), pp. 221-229.

behaviors. As our students are going out and will spend several decades in industry, our concern is about the future of CAE and how to prepare them to retain competency throughout their whole careers. So this paper will report on the history of FEM and a prediction for the future of CAE. Naturally, we will discuss the importance of systematic innovation tools and our experiences with implementing this into engineering design class and feasibility of engineering curricula reformation are introduced. So the first part will examine current practice of CAE and the needs of systematic innovation tools. The second part describes roles of CAI in engineering design classes.

## 2 Advancement of CAE

### 2.1 History of FEM

Considerable achievements have been made to avoid catastrophic failures, although there have been some exceptions such as the 2003 Columbia accident [4]. And one of the achievements of CAE is to make robust estimation of durability. CAE is critical element in VPD (Virtual Produce Development) in conjunction with current digital enterprise including ERP, PLM, and SCM. The core of current CAE practice is FEM as shown in Fig. 1.

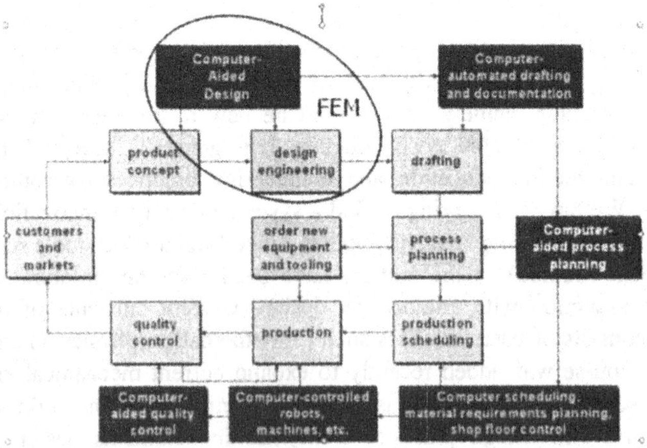


**Fig. 1.** FEM in CAE processes

The first part of this section investigates the evolution of FEM. Simulations of product behavior are necessary for good design. Mostly the governing equations are given as differential equations and the final problems become BVP (Boundary Value Problems) or IVP (Initial Value Problems). Analytical solutions can be obtained for simple geometries with simple conditions and exhaustive solutions are documented for many cases. Many eminent applied mathematicians and engineers contributed on the exact solutions, but most real problems do not allow analytical solutions. Therefore we should resort to numerical methods such as FDM (Finite Difference Method) and FEM. The discretization methods are developed for this purpose and FDM has longer history than FEM in this sense. But FEM plays more important role as the de facto standard method nowadays. FEM uses 'Divide and Conquer' concept and discretizing techniques. The FEM has developed into current form by combining variational and weighted residual method [5,6]. In the beginning, continuous trial functions were applied to discrete structures such as trusses and extended to continuum objects later. The modern concept of FEM was initiated in the 1950's while the implementation was done in the 1960's. Core concepts were ripe during the 1970's and extended to every possible problem after digital computers were introduced and increased their capacitances. Naturally, FEM was applied to structural analysis first and the three dimensional formulation was established. Ironically the solution process is easier for 3D problem than 2D problems such as plates and shells. The numerical techniques for these thin structures are still hot research topics. Next FEM solved more PDE's which govern stress and thermal problems. It is quite surprising that the evolution of FEM followed exactly the technology evolution pattern suggested by TRIZ. TRIZ is not explained here and readers can find information about TRIZ easily [7]. This pattern is summarized as MeThAChEM – which means Mechanical, Thermal, Acoustical, Chemical, Electrical, Magnetical and Electromagnetical. Those solutions are the first phase of MeThAChEM, namely mechanical problems. Usually PDE's are categorized into elliptic, parabolic, and hyperbolic types. The stress analysis problems are elliptic and the thermal problems are parabolic. The third is hyperbolic problems and the representative problem is a wave propagation problem. Mathematical solutions had long histories for this type of problem, but practical application of FEM to this kind appeared around 1970's which was much later than the former two types [8]. The solutions of wave phenomena are A (Acoustical) in MeThAChEM. Next applications are on chemical problems where the mass transfer is the most important distinction between mechanical and chemical analysis. The practical finite element solutions appeared on mid 1980's long after the mechanical solutions were obtained. Recently electro-magnetic problems have been analyzed by finite element method. It is wondrous to the author that the development of finite element and application procedures followed the basic concepts of TRIZ even though it had been developed independently for long time. The FEM started from dealing with matters in terms of continuum but now it is used to analyze field problems. Microscopically FEM was developed to handle discrete structures such as trusses. It was extended to deal with continuum later. It is ironic that FEM is now struggling to successfully combine with quantum mechanics which has a discontinuous nature. Those researches are being done in the realm of meso-mechanics. Although there is much literatures on the history of FEM, it is the author's opinion that this is the first attempt to combine FEM and TRIZ.

## 2.2 Future of CAE

Naturally the next question is to predict the future of CAE where FEM represents a core element. This general pattern is also quite similar to the direction of ideality of TRIZ, which means the output from input goes infinity as the system evolves. CAE will also follow the incremental pattern of ideality and the future can be described as follows. The hardwares of computing will give following potentials at around 2020. Typical consumer computers will have multiprocessors, 3 Tera Hz cpu, ½ TB of RAM and 50 TB of disk space. And the sensors will be cheap and ubiquitous [9]. So the engineering analysts used to fulfill the role of programmers and could now be categorized as analysts should be users and operators of these facilities in the future. For the software side, the HCI (Human computer interface) will improve consistently until we reach the stage where every engineer can perform any simulation he wants to do without difficulty. In our laboratory, we are using ABAQUS for FE analysis and HyperMesh for pre-post processing. Both analysis and pre-post processing softwares are improving very much for every version upgrade. Besides, COMSOL claims it can provide multi-physics capabilities inherent in its software. Also many symbolic computation softwares enable scientists and engineers to simulate whatever they want to do. They include MATHEMATICA, MACSYMA, MAPLE and MATLAB. Recent report on the research direction in computational mechanics, where the main tool is finite element, selected the following items as the main themes to dig into in the future: Virtual design, multi-scale modeling, adaptivity, very large-scale parallel computing, biomedical engineering and controlling uncertainty [10]. One of the important problems is to consider uncertainty in design parameters. There are 3 kinds of uncertainties, namely material property variability (Fatigue), manufactured shape variability (Tolerance) and design loads variability (Environmental). Almost all deterministic problems have been solved and the probabilistic problems are now main research topics. When we follow the trend of the problem types and computational power, the evolution of FEM may be summarized as follows.

- 1970's
  - Linear statics, Buckling, Modal analysis
  - 5,000 grids, 3,000 elements, 25,000 DOF
  - 2,000 CPU sec, 1 Elapsed hr, 8MB memory, 300 MB disk
- 1980's
  - Nonlinear, Dynamics, Superelements, Direct methods
  - 50,000 Grids, 60,000 Elements, 250,000 DOF
  - 2,000 CPU sec, 1 Elapsed hr, 480 MB memory, 1GB disk
- 1990's
  - Adaptivitiy, Optimization, Coupled analysis
  - 150,000 Grids, 200,000 Elements, 1.5M DOF,
  - 6,500 CPU sec, 2 Elapsed hr, 1GB memory, 100GB disk
- 2000's
  - Probabilistic, CAD integration, AI
  - No limit (As hardware allows)

The relationship between TRIZ and CAE or FEM is investigated from the history of FEM. The first applications of primitive form of FEM were on discrete structures and extended to continuum problems later. This process can be considered as an example of 6th rule among evolution patterns of technical systems in TRIZ, increasing complexity, and 4th rule, increasing dynamism. And the development of modern FEM follows exactly the same patterns of MeThAChEM as discussed. Although the origin and main concepts are mathematical and computational, the patterns of evolution of the FEM are amazingly same as those used by TRIZ. So the future of CAE or FEM can be predicted by extending the line of evolution pattern. Certainly CAE will advance in the direction of increasing ideality. From the research items in the report [10], the patterns of TRIZ can be matched as follows.

Virtual design : As the system evolves increasing ideality, everything could be designed using field concept instead of substances. So the CAE systems will be virtual.

Multi-scale modeling : So far FEM is used for continuum level simulation. But the capability of FEM will be on meso and quantum level. Naturally multi-scale modeling will be an essential feature of future FEM. This is exactly same as 7$^{th}$ rule of the TRIZ pattern, transition from macro to micro.

Adaptivity : This can be matched with 2$^{nd}$ rule, increasing ideality, 4$^{th}$ rule, increasing controllability and 8$^{th}$ rule, increasing automation.

Very large-scale parallel computing : There are two ways to handle big computational tasks. One is to develop a faster chip and build so-called supercomputers. But this trend has inherent limitation and increases the expenditure very much. The other approach is to use parallel systems. This method is good to make super computing power using normal pc's. Also it is well known that new projects are developed to use idle computers to solve big problems which can not be solved by one super computer. World Community Grid is one example of that kind of effort [11]. This matches well with the main concept of TRIZ, which emphasizes 'use of every possible resource'.

Biomedical engineering : The other big area for future engineering application is BME(Biomedical engineering). Recently bionics and TRIZ were combined together for creative design methodology. One of the future challenges of FEM is also for exploration of living systems.

Controlling uncertainty : This can be matched with 8$^{th}$ rule, decreasing human involvement or increasing automation.

When we apply IFR(Ideal Final Results) of TRIZ to the computer systems, the final computing practices will be done without expensive hardwares or system softwares. Every engineer should be able to do realistic simulations without profound knowledge on finite element theory or computing systems and programming languages. My graduate students are working at CAE teams of big companies showing their expertise on FEM. But soon they should demonstrate another capability of creativity when FEM becomes a routine engineering task. So far there is a specialty called 'numerical analyst', but the name will soon disappear. That is also the reason that students of engineering science should learn TRIZ more seriously. The ideal final CAE systems will free engineers from solutions of difficult differential equations and long hours to master FEM proficiently. By then the engineers can concentrate only on creative and innovative idea generation [12].

Engineering design can be divided into formative and creative design. The formative design is incremental design, routine design, improvement, optimization, and compromise. Creative design is innovative design that aims to implement an original concept beyond current paradigm. Even though the importance of creative design is recognized, there are no other methods than TRIZ to enhance systematic creativity in real world design problems.

# 3 Implications on Engineering Education

## 3.1 New required capabilities of engineers in global design environment

For an engineering educator, the required skills which are needed as global engineers are the most important concern. Hayes and Pande identified 5 trends in the global market place as more internationally distributed organizations, more distributed product development teams, more complex products, more products aiming at emerging markets, and products needed to be more usable and appealing [13]. They also summarized 5 more skills in addition to technical expertise for effective global engineers including cross-cultural skills, systems thinking skills, willingness to work outside one's own discipline, design for usability skills, and new approaches to design management. As a structural mechanics professor studying CAE mainly on FEM, I feel that we should add creativity to achieve those goals. Next section will examine this idea.

## 3.2 Systematic Innovation

One of the big concerns of practicing engineers is that these capabilities do not guarantee a commercial success of an engineering product. Some technological breakthroughs are required in addition to fancy colorful drafts of analysis results and designs. So a new paradigm for the product development is sought and CAI (Computer Aided Innovation) can be the answer for that requirement. Naturally CAI, such as TRIZ softwares are rapidly gaining momentum for practical applications. As confessed in the article of ASEE (American Society of Engineering Education) magazine, PRISM, the engineering education community couldn't find good tools for creativity except serendipity [14]. Only luck and probability dominates in this much laborious approach. The article pointed out TRIZ as the potential solution for creativity in engineering education. Prof. Flowers of MIT stressed that creativity is so important that 'informed creative thinking' must be taught in engineering education in his plenary talk at the 2004 ASEE conference [15]. The reason is that smart students can be obtained by outsourcing but the leadership can not be outsourced. In addition to that point he declared engineering science is dead and real engineering is needed in engineering education. There are many debates on what creativity is and how to foster creativity. Researches on creativity can be divided into 3 categories, namely on creative people, methods and products. One Japanese scholar summarized 88 methods of creativity [16]. But the important thing is how to make people learn and improve their creative activities. So the systematic innovation tools are long sought and developments into softwares are now in the realm of CAI. We spent a long time to investigate these tools and found TRIZ could be much helpful for our education and also for generating new products and patents.

### 3.3 CAI in Engineering Education

In our school, TRIZ has been introduced to the class for 7 years in the courses of Creative Problem Solving (CPS) for freshman, Creative Engineering Design (CED) for junior and New Product Design (NPD) for senior. Traditional engineering design texts have been used for many years before TRIZ since Engineering Design courses have been a mandatory course. Recent Korean texts are being used for design education [17,18]. The first year engineering students practice engineering process of design using LEGO system and the concept of creative practice is just explained. At the junior course, a more serious process is adopted. Teams are made of 4 students to make as diverse as possible. They are assigned 2 projects during the semester. The first is targeted to bring 'Eyes for Design'. They should make a long list of items that they feel represent bad or uncomfortable design. This part is 'Need Finding' stage, which may be considered as same as QFD in product development. Next project is to improve or solve that situation. Usual outcomes from this course for many years before introduction of TRIZ were CAD drawings. At most better reports included the FE analysis results. Virtually no outcomes were considered to have serious practical meaning. After the introduction of TRIZ, the results became several applications of patents and designs at the end of semester. Some of them were developed into more serious government funded projects. TRIZ contributed by motivating the students to follow all of the steps of patent applications and also give confidence in the sense of what they can do. From last year, we started to offer senior NPD course by popular requests from students. Here we directly apply TRIZ principles and processes for given subject by using 3 commercial TRIZ softwares, namely Goldfire, Innovation WorkBench, and CREAX. Their responses and the outcomes are great. Every team prepared patents and/or business plan for their products. And they made a study group of TRIZ and became evangelists among students. I have not seen such enthusiastic responses and practical outputs before. This fact gives a big momentum for Korean universities and there is a movement to form a coalition for creative engineering design adopting this approach. This can be a manifest of the trend 'from engineering science to engineering'. It also gives 2 implications to engineering education. As the required credits for a degree become smaller in current college of engineering curricula, the material of the basic science courses should be compacted and more courses of exercising creativity should be developed. In the light of this trend, the purpose of engineering education involves from training to adapt current technology to breeding self-learning capabilities. We found the approach of 'Effects' module in Goldfire is very promising. They collected 8000 important scientific principles and categorized by verbs applied to fields, parameters and substances. This taxonomy of engineering activities can be more elaborated and transferred to engineering students efficiently. One of the important topics in solid mechanics which is one of the core courses in mechanical engineering is to analyze a cantilever beam. Only few students recognize the usage of a cantilever beam as a sensor on the airbag until they were asked about the usage. Instead of extending 1-d mechanics to bending problem, we may ask them to find ways to measure the acceleration. As this kind of extension has big potential for engineering education, we organized a multi-university research team and started a project to elaborate the taxonomy of engineering activities. This approach is expected not only to boost student's problem solving capabilities overcoming narrow sight in a single core discipline but also to give practically more freedom to concentrate on the essence of problems instead of on the methods themselves. These efforts are applied in diverse ways of teaching to utilize current IT advancements. Distance learning and certification of TRIZ is being tried for reeducation of practicing engineers. This system is brought into AUCSE (Ajou University Consortium for Staff Enrichment) which was built in 2006 to educate the people and solve the problems of small and medium sized companies. The most widely sought topics from survey were some depth technology of their own and the capabilities to produce successful new product. More elaborate study on the education system of these approaches is expected to change current engineering education system.

# 4 Conclusions

The vision of next-generation engineers may be described as follows. Every engineer with capability of simulating whatever they want to do using computer and IT technology is concentrating on his mission and solve by exercising his creativity by help from CAI softwares. To this end, we have examined the history of FEM and predicted the future of CAE. The necessity of creativity came out naturally from this investigation and our experiences of implementing TRIZ into our design classes are described. The efforts to renovate current engineering education systems are also introduced. All attempts to collaborate in this direction and globalize engineering education are welcomed. And this is the point where we should show our learned creativity and exercise CAI capabilities.

# References

1. L. Leifer, et. al., *Radical Innovation: How Mature Companies Can Outsmart Upstarts* (Harvard Business School Press, 2000).
2. M. Stefik, and B. Stefik, *Breakthrough : Stories and Strategies of Radical Innovation* (The MIT Press, 2004)
3. W. Sun, Strategy to Connect Manpower of Industries and Universities, Mechanical Engineer's Day Conference, KSME (2004)
4. CAIB (Columbia Accident Investigation Board) report, USA http://caib.nasa.gov (2003)
5. O.C. Zienckiwicz, R.L. Taylor, and J.Z.N. Zhu, *The Finite Element Method*, (Butterworth-Heinemann, $6^{th}$ ed., 2005)

6. C. F. Williamson, Jr., A History of the Finite Element Method to the Middle of 1960's, Ph. D. Thesis (Boston University, 1976)
7. Homepage of the TRIZ journal, http://www.triz-journal.com/
8. P. Bettess, and O.C. Zienckiwicz, Diffraction and Refraction of Surface Waves Using Finite and Infinite Elements, Int. J. Numer. Meth. Eng. **11**, 1271-1296 (1977)
9. M. Bryden, Virtual Engineering : Playing the Real Game, Seminar at Univ. of Maryland (2007)
10. J.T. Oden, et. al., Research Directions in Computational Mechanics, Comp. Meth. in Appl. Mech. and Eng., 192, 913-922 (2003)
11. World Community Grid Hompage : http://www.worldcommunitygrid.org/
12. N. Leon, and O. Martinez, Product Optimization vs. Innovation, Steps toward a Computer Aided Inventing Environment, Proc. TRIZCON 2003, Philadelphia, PA, 2-1 ~ 2-12 (2003)
13. C. C. Hayes, and A. Pande, What Skills Will Future Engineering Graduates Need in Global Organizations?, Proc. Int. Design Eng. Tech. Conf., ASME (2007)
14. D. McGraw, Expanding the Mind, PRISM, ASEE, **13**(9), 30-36 (2004)
15. W. Flowers, Plenary Talk, ASEE Annual Conference, June 27 (2004)
16. T. Makoto, *The Bible of Creativity* (Moodo School Press, Japan, 1993) *(In Japanese)*
17. S.H. Yoo, *Designer's Creativity*, (Ajou University Press, Suwon, Korea, 2004) *(In Korean)*
18. I.C. Kim, *What to make?*, (Intervision, Korea 2006) *(In Korean)*